Roland Kankey Jane Robbins
Editors

Cost Analysis and Estimating

Shifting U.S. Priorities

With 34 Figures

Springer-Verlag
New York Berlin Heidelberg London
Paris Tokyo Hong Kong Barcelona

Roland Kankey
Jane Robbins
Air Force Institute of Technology
Wright Patterson Air Force Base
Ohio 45433
USA

Printed on acid-free paper.

Camera-ready text prepared by the authors.

9 8 7 6 5 4 3 2 1

ISBN-13:978-1-4612-7831-3 e-ISBN-13:978-1-4612-3202-5

DOI:10.1007/978-1-4612-3202-5

FOREWORD

The environment for today's cost estimator and analyst is certainly very challenging. Computerization, software, robots, composites, uncertainty, and integrated systems all challenge the applicability of our existing tools and techniques. These *Proceedings* serve to document some of the completed and on-going research in the dynamic world of costing.

This document is published in conjunction with the first Society of Cost Estimating and Analysis (SCEA) National Conference, held in Boston, MA, June 19–21, 1991. It serves to foster and promote cost research, and to provide a forum to report these findings in furtherance of public interest.

This volume is the third of the series. The first and second were published in conjunction with the 1989 ICA/NES Joint Conference in Washington, D.C., and the 1990 ICA/NES Joint Conference in Los Angeles.

My thanks to our Editors, Professor Jane Robbins and Dr. Roland Kankey; our Managing Editor, Mr. Frank Flett; the Program Chair, Ms. Ann-Marie Sweet; and all those who contributed.

R. R. Crum, President
Society of Cost Estimating and Analysis

PREFACE

We wish to thank the professionals who submitted papers to us for review. As any editor will indicate, you cannot review or publish papers that are not submitted.

The articles in this *Proceedings* successfully completed the referee process. Each of these authors was rewarded by an additional cycle of minor changes, word processing, and express mailings.

We are deeply indebted to those individuals who supported us. Among the individuals who provided support were Ms. Ellen Burton who prepared all the correspondence to the authors; Ms. MaryEllen Ogle who did extensive proof–reading of the manuscripts; Ms. Patricia Pietrzak who served as our special administrator and coordinator; Ms. Gail Elm who did quality assurance; Mrs. Denise Farmer who spent endless hours typesetting the manuscripts; and Mr. Robert Hazlett and Mr. J.D. Weaver. We would also like to thank the people at the Air Force Institute of Technology's School of Systems and Logistics who supported work on this volume; specifically, we thank Col Richard S Cammarota; Dean, and Dr. William Mauer, Associate Dean; LtCol Curtis Cook, Head of the Department of System Acquisition Management; Mr. Leo Mercier, former Head of the Department of Quantitative Management; and Prof. William Dean, Deputy Department Head, Department of System Acquisition Management. Many others also assisted and without their help, this volume would not have been possible. In particular, the support of the Dayton office of TASC is gratefully acknowledged.

We are also deeply indebted to the following individuals who served as referees for the complex task of reviewing manuscripts.

Scott N. Belford
William Bowlin
David S. Christensen
Daniel Ferens
Paul R. Garvey
Jack Gibson
Thomas R. Gulledge Jr.
Earl King
Steve Klipfel
Roger Koble
Ralph Liebhaber
John A. Long, Jr.
Wayne S. Maricle
Gerald McNichols
Richard Murphy
William Richardson
Thomas Schuppe
M. Dale Shields
John Tarter

CONTENTS

A FRAMEWORK FOR ESTIMATING THE COST TO BUILD TRUSTED COMPUTER SYSTEMS

Paul R. Garvey

The MITRE Corporation
Bedford, Massachusetts 01730

ABSTRACT

Advances in the field of computer network technology have allowed unprecedented levels of information sharing to be possible between users. As a result, there has been an increased need within the Department of Defense (DOD) to protect sensitive information and data sources against unauthorized access or disclosure. Reflective of this, the number of USAF Electronic Systems Division (ESD) acquisitions being directed to meet stringent security requirements is increasing. Absent from the cost analysis community has been a systematic approach for estimating the resources to build computer systems that are trusted to protect the information they process.

This paper presents a framework, based on research conducted by The MITRE Corporation, that identifies and schedules the security engineering tasks necessary to build a trusted computer system. For the program manager, this framework provides for the technical planning of security-relevant engineering activities around an acquisition's major development milestones. The framework has been constructed in sufficient detail to support level of effort costing and, as a consequence, permits the cost analysis process to be directly incorporated into a project's specific security engineering approach. A database has been initiated on security engineering costs, and insights into the major cost drivers associated with specific security requirements are shown.

The security engineering task schedules and the effort data presented in this paper, collectively provide an approach for estimating the cost to build trusted computer systems that meet DOD 5200.28-STD requirements. This research represents our first step in evolving a cost methodology sensitive to the very complex system-wide technical issues involved with building secure systems for the DOD. We offer the research summary contained in this paper, so that the defense cost and security technical communities may review, comment on, and expand upon the approach within their organizations.

1. EXECUTIVE SUMMARY

There is an increasing need within the Department of Defense (DOD) to require the development of hardware and software mechanisms to protect sensitive data contained in automated information systems (AIS). Providing mechanisms that enforce strict access policies in systems that interface to a broad user community, with varying levels of clearance and need–to–know, is among the most challenging problems presently confronting information technology.

In 1985 the Department of Defense issued a standard for performing technical security evaluations of automated information systems. It is known as the Trusted Computer System Evaluation Criteria (TCSEC) and published as DOD 5200.28–STD [1]. According to the system's security class, the TCSEC provides the minimum set of security requirements that must be engineered into an evolving AIS. Revised in March 1988, the TCSEC has served, and presently remains, the only officially approved standard for evaluating the degree of protection provided in a trusted computer system.

In the years since DOD 5200.28–STD was issued, there has been little research into the cost associated with building trusted computer systems. In response to this, MITRE has undertaken to evolve methodologies for estimating the cost to build trusted computer systems through two cost research efforts. One is being conducted by the Economic Analysis Center at MITRE–Washington. They have been investigating the effort to build trusted code, and are examining the feasibility of calibrating commercial software resource models to estimate the cost to develop trusted applications software. The research presented in this paper was conducted by the Cost Analysis Technical Center at MITRE–Bedford, for the purpose of achieving two goals. First, we sought to develop a technique that would identify the areas where security requirements have a cost impact across the system. Second, determine what is driving these impacted areas and make some estimates on the magnitude of the cost. In summary, the research presented herein, has focused on the security engineering process involved with building trusted computer systems, presents a detailed assessment of how specific security requirements and tasks impact the system and software development, and presents effort data collected from programs that have, or are in the process of, implementing the TCSEC requirements of DOD 5200.28–STD.

A summary of the key findings yielded by this research are presented below.

- The requirement for a trusted computer system increases the engineering development costs not just in prime mission software engineering (and possibly hardware), but also in the support cost elements of system engineering, system test and evaluation, data, and operational site activation.
- Contractor cost reporting practices on programs that follow MIL–STD–881A [2] guidelines do not require a separate line item for security engineering costs. In such cases, it is almost impossible to completely extract the cost for security engineering, which is blended with the costs for system engineering, system test and evaluation, data, and configuration management. Furthermore, under fixed–price contracts, there is no requirement for a contractor to establish the special accounting mechanisms that may be needed to provide insight at the level where the security engineering cost impact can be accurately viewed.
- Security requirements must be considered in all engineering design decisions that take place during the early system design reviews. Attempts to retrofit security into an existing architecture can be costly and have often failed, and in many instances the security functionality originally targeted had to be down–scoped to remain within the system performance limitations and/or development schedule constraints.

Three key points are presented for the reader's attention.

- The general–purpose secure operating systems, which presently exist on the evaluated products list, are too limited in type and generally do not meet the combined security and mission requirements (e.g., fault–tolerance, performance) of systems being acquired by the DOD. This situation is not expected to substantially improve in the near–term. Until a more expanded set of commercial products becomes available, acquisitions will continue to face additional costs as contractors either modify existing commercial products or develop mission–unique products to meet the operational and security requirements of today's systems.
- The TCSEC is the only official DOD standard in the computer security field. Furthermore the TCSEC was written for the evaluation of general purpose operating systems, and the requirements it contains must be extrapolated and interpreted when applying the criteria to developing a trusted application system (e.g., one where users interact with the system through applications software, rather than using the system's general purpose capabilities). Because of the mission–specific character of application systems, no general purpose

"trusted application system interpretation" is being written. Interpretation of the TCSEC for application systems is done on a system-by-system basis.

- The security of sensitive information processed by a trusted computer system ultimately rests on the integrity of its users. Even the most sophisticated security engineering designs cannot guard against unlawful tampering by a legitimate user.

Table 1-1 presents a statistical summary of the effort data, shown in staff months, collected by the industry survey phase of this study. Surveys were conducted across five companies to obtain effort data reflective of the security requirements found in the B1 and A1 security classes of DOD 5200.28-STD. A detailed summary and analysis of the leading security engineering task drivers is the topic of Section 4.

The information presented in Table 1-1 reflects the system-level efforts associated with the support cost elements of MIL-STD-881A. Thus, these efforts are in addition to likely increases in prime mission software and/or prime mission hardware development costs, which should be estimated on a project-specific basis.

Table 1-1 Survey Data Summary (Staff Months)

SECURITY CLASS	GOVERNMENT & CONTRACTOR TASK	CONTRACTOR TASKS ONLY
B1		
Mean	166	132
Standard Deviation	26	24
A1		
Mean	348	303
Standard Deviation	64	62

Security cost and technical data were gathered on three ESD projects. Two of these are ongoing embedded Command, Control, and Communications (C^3) system development projects and the other is a minicomputer based system that will provide a multi-level secure interface between two networks.

The two ESD embedded C^3 system development projects are targeting for the B2 security class and implementing the multilevel operational mode. Interestingly, when they are compared, their combined efforts for system security engineering design and support to system test and evaluation (these are considered to be non–Prime Mission Product (PMP) tasks) are nearly the same at approximately 30,000 staff hours, which is approximately 200 staff months. In addition to this effort, one of these two projects revealed that an additional 36,000 source lines of code (approximately 21 percent of the total software size) was being developed to achieve the security–related requirements.

The third ESD project is a minicomputer based system that will provide a multi–level secure interface between two networks. It is not an embedded C3 system that must incorporate security requirements into an overall set of mission requirements. Security engineering is the primary focus on this project and most of the engineering activities could be categorized as PMP tasks. The overall effort estimated to complete the project, including the software development, is on the order of 30,000 staff hours.

The amount of historical data collected is too scant to support the rigorous development of security engineering cost equations based on an underlying economic law or statistical analysis, however, the information gathered does suggest the following general indications:

- There can be a significant amount of non–PMP effort for security engineering design and analysis and security engineering support to system test. The industry survey indicated that this effort could range from 132 to 303 staff months. In the case of the two ESD embedded C^3 system development projects this effort is approximately 200 staff months;
- There is an impact on the amount of code, the software element of PMP, developed to implement the security requirements of the system. On one of the C3 projects, current estimates indicate that 21 percent (approximately 36,000 source lines of code) of the total amount of code developed is due to the security requirements. On the third ESD project, an estimated total of 38,000 lines of C code are being developed to implement a multi–level secure interface.

To conclude, Eq. 1 offers a heuristic formulation of the system–wide cost for security. The equation is not intended as a basis for parametric cost estimation; rather, it identifies a functional relationship between the elements of a MIL–STD–881A work breakdown structure that must be estimated, when developing the system–wide cost of security requirements.

where

$$\text{Security Cost} = \Sigma\ \text{nonPMP} + \alpha \frac{I_{sec}}{P_r} + \Sigma\ \text{PMP HW} \tag{1}$$

ΣnonPMP	=	the sum over all non–PMP costs impacted by security requirements
α	=	a software development labor rate (dollars per staff month)
I_{sec}	=	the estimated lines of code attributable to the security requirements
P_r	=	the software development productivity rate (lines of code per staff month) of the project
ΣPMP HW	=	the sum over all PMP hardware costs impacted by the security requirements

2. RESEARCH SCOPE

This research represents the results of a five month effort to investigate the cost of building trusted computer systems. The goal of this research was to improve our understanding of the elements which contribute to the cost of building secure systems, and to develop a framework around these elements that provide a systematic cost estimation approach and technical planning tool. To achieve this, industry surveys were conducted to determine the nature and scope of the security engineering tasks and the efforts associated with the development of a trusted AIS. Programs implementing TCSEC requirements were studied to construct a detailed view of the security engineering process and activity hierarchy necessary to build and accredit a trusted computer system.

Prior to this study, we had limited insight into the impact that security requirements could have on the system development costs. We estimated security costs by increasing the size of the development software and by increasing the level of effort for system test. However, a detailed examination of the engineering approaches that contractors employ to implement TCSEC requirements, revealed how extensively security engineering effort spans across many work breakdown structure (WBS) elements of a system. Thus, adjusting cost estimates for the impact of security on just the prime mission software and on the system test cost element understates the total system engineering effort. Therefore, the scope of this research was to determine, more completely, the overall PMP and system level (the non–PMP elements of MIL–STD–881A) cost impacts on an AIS implementing TCSEC security requirements.

The following section presents an overview of the evaluation classes described in the TCSEC and the relationship between the TCSEC security requirements and a system's protection level. This sets the stage for the development of a generalized security engineering task schedule, which is presented in Section 3.1. The generalized security engineering task schedule provides a framework suitable for developing level of effort estimates of the cost to build trusted computer systems.

3. AN OVERVIEW OF THE EVALUATION CLASSES

The Trusted Computer System Evaluation Criteria, also referred to as the 'Orange Book', provides a standard for specifying security requirements in acquisition specifications, establishes guidelines to manufacturers regarding the security features to build into commercial products, and presents a procedure for evaluating the degree of trust that can be placed in computer systems for the secure processing of classified and other sensitive information.

The criteria in the Orange Book are ordered hierarchically and divided into the D, C, B, and A divisions. Moving from division D to division A implies major improvements in the system's ability to protect sensitive information. Division A provides the most comprehensive security. Within divisions C and B there are subdivisions known as classes.

Table 3–1 presents the relationship among protection level, division, and evaluation class. The evaluation classes are shown as a function of the division and the protection levels. The TCSEC requirements are additive and increase in functionality and assurance as one ascends the evaluation classes.

The following will present a brief summary of the technical complexity associated with each protection level, and how it increases dramatically from minimal protection (division D) to a verified design (division A). Sources of information regarding the target evaluation classes for products described below originated from the National Computer Security Center (NCSC) product bulletins and the vendors.

Division D provides minimal protection, and it contains only one class. This division is reserved for those systems that have been evaluated, but fail to meet the security requirements for a higher evaluation class.

Table 3-1 The Evaluation Classes

PROTECTION LEVEL	DIVISION	EVALUATION CLASS
Minimal Protection	D	D
Discretionary	C	
Discretionary Security		C1
Controlled Access		C2
Mandatory	B	
Labeled Security		B1
Structured Security		B2
Security Domains		B3
Verified	A	
Verified Design		A1

Division C provides discretionary (need-to-know) protection. The C1 and C2 classes in this division provide for discretionary protection and accountability of users and, in class C2, the actions they initiate through the inclusion of audit capabilities.

The security-relevant portions of a system, referred to as the Trusted Computing Base (TCB), of class C1 systems nominally satisfy the discretionary security requirements by providing separation of users and data. The class C1 environment is expected to be one of cooperating users processing data at the same level(s) of sensitivity. Most of the major commercial operating systems offered today have incorporated class C1 requirements.

Systems in class C2 enforce a more finely grained discretionary access control than class C1 systems by making users individually accountable for their actions through login procedures and through the auditing of all security-relevant events. An example of this class is Digital Equipment Corporation's VAX/VMS general purpose operating system Version 4.3.

Division B provides mandatory protection. In this division the TCB preserves the integrity of sensitivity labels and uses them to enforce a set of mandatory access control rules.

Class B1 systems require all the security features for class C2. In addition, an informal statement of the security policy model, data labeling, and mandatory access control over named subjects and objects must be present. A subject is an active entity generally in the form of a person, process, or device that causes information to flow among objects or changes the system state. An object (e.g., files, programs, words, bits/bytes) is a passive entity that contains or receives information. Mandatory access control, in class B1, is a means of restricting access to objects based on the sensitivity (as represented by a label) of the information contained in the objects and the formal authorization (clearance) of subjects to access information of such sensitivity. In class B1 the capability must exist for accurately labeling exported information, and all flaws discovered during testing must be removed. Two vendors, AT&T and UNISYS have their respective products VMLS and OS1100 on the EPL.

In class B2 systems the TCB is based on a clearly defined and documented formal security policy model that requires the discretionary and mandatory access control enforcement found in B1 systems to be extended to all subjects and objects of the system. In addition, covert channels are addressed. The TCB must be structured into protection–critical and nonprotection–critical elements. In class B2, the TCB interface is well defined and its design and implementation enable more thorough testing and review. Authentication mechanisms are strengthened, trusted facility management is provided in the form of support for systems administrator and operator functions, and stringent configuration management controls are imposed. Class B2 systems are relatively resistant to penetration. The only general purpose operating system on the EPL for class B2 is Honeywell's Multics MR11.0 system. Presently, Trusted Information Systems (TIS) and Verdix are targeting their respective products Secure Xenix and VSLAN for formal class B2 evaluation.

Security domains is the final class in division B. The class B3 TCB is required to mediate all accesses of subject to objects, be tamperproof, and be small enough to be subjected to analysis and tests. Thus, the TCB is structured to exclude code not essential to security policy enforcement, with significant software engineering effort expended during TCB design and implementation directed toward minimizing its complexity. A security administrator is supported, audit mechanisms are expanded to signal security–relevant events, and system recovery procedures are required. Class B3 systems are highly resistant to penetration. An example of a candidate system for class B3 is the Gemini Multiprocessing Secure Operating System (GEMSOS) designed to the class B3 security requirements for communication and workstation applications.

Verified protection, division A, is characterized by the use of formal security verification methods to assure that the mandatory and discretionary security controls employed in the system can effectively protect the classified and other sensitive information stored or processed by the system. Extensive documentation is required to demonstrate that the TCB meets the security requirements in all aspects of design, development, and implementation. Systems in class A1 are functionally equivalent to those in class B3 in that no additional architectural features or policy requirements have been added. The distinguishing feature of class A1 systems is the analysis derived from formal design specification and verification techniques, which results in a high degree of assurance that the TCB is correctly implemented. This assurance is developmental in nature starting with a formal model of the security policy and a formal top-level specification (FTLS) of the design. More stringent configuration management is required and procedures are established for securely distributing the system to sites. An example of this level is the Honeywell Scomp 2.1 general purpose operating system. Presently, Boeing is targeting their multi-level secure (MLS) local area network (LAN) product for formal class A1 evaluation.

Table 3–2 [1] presents the relationships among a system's security requirements areas, its division classification, and its evaluation class. In many areas, the requirements for a given evaluation class are more stringent than those for lower classes. For example, all evaluation classes have requirements in the area of discretionary access control; however, the requirement for class B3 is considerably more stringent and difficult to meet than those for class C1.

The security requirements listed in Table 3–2 reflect a general set of minimum requirements for each of the six evaluation classes defined in DOD 5200.28–STD. As the number of requirements increase across these classes, so does the technical complexity and the associated cost of the security engineering. Note that the largest single increase in the minimum number of requirements occurs between C2 and B1. This is because B1 is the first class to implement the labeling and mandatory access control requirements of a security policy, which accounts for seven additional requirements over C2. The last requirement distinction between these two classes is in assurance. Class B1 is the first to require, under the design specification and verification assurance requirement, that a model of the security policy be demonstrated to be consistent with the policy's axioms. Although it is not obvious from Table 3–2, there is a significant engineering level–of–effort jump between class B1 and class B2 due to the system architecture requirement and not due just to the additional functionality.

3.1 THE GENERALIZED SECURITY ENGINEERING TASK SCHEDULE

A generalized security engineering task schedule has been developed across all six TCSEC classes described in section 3.0. Links have been made on the schedule between the security engineering tasks and their relationship to the evaluation classes defined in the TCSEC. This schedule is generally representative of the set of security engineering tasks that are typically

Table 3–2 Security Requirements Versus Evaluation Class

REQUIREMENT	EVALUATION CLASS					
	C1	C2	B1	B2	B3	A1
SECURITY POLICY						
Discretionary Access Control	*	*	*	*	*	*
Object Reuse		*	*	*	*	*
Labels			*	*	*	*
Label Integrity			*	*	*	*
Exportation of Labeled Information			*	*	*	*
Exportation to Multilevel Devices			*	*	*	*
Exportation to Single–level Devices			*	*	*	*
Labeling Human–Readable Output			*	*	*	*
Mandatory Access Control			*	*	*	*
Subject Sensitivity Labels				*	*	*
Device Labels				*	*	*
ACCOUNTABILITY						
Identification & Authentication	*	*	*	*	*	*
Audit		*	*	*	*	*
Trusted Path				*	*	*
ASSURANCE						
System Architecture	*	*	*	*	*	*
System Integrity	*	*	*	*	*	*
Security Testing	*	*	*	*	*	*
Design Specification and Verification			*	*	*	*
Covert Channel Analysis				*	*	*
Trusted Facility Management				*	*	*
Configuration Management				*	*	*
Trusted Recovery					*	*
Trusted Distribution						*
DOCUMENTATION						
Security Features Users Guide	*	*	*	*	*	*
Trusted Facility Manual	*	*	*	*	*	*
Test Documentation	*	*	*	*	*	*
Design Documentation	*	*	*	*	*	*
Total Number of Requirements	9	11	19	25	26	27

required in the acquisition of a trusted computer system. From a cost perspective, the schedule provides an acquisition architecture, or framework, suitable for deriving level–of–effort (LOE) estimates against the tasks identified on the schedule.

The generalized schedule reflects a set of tasks typical for inclusion in the security–relevant portions of a program's statement of work (SOW). The Trusted Computer System Evaluation Criteria classes previously described, presents the set of minimum security requirements for the various levels of trust that may be required of a system. Figure 3–1 provides a mapping that shows which SOW tasks satisfy the minimum set of requirements defined in the Trusted Computer System Evaluation Criteria.

The SOW tasks shown in Fig. 3–1 reflect a general set of engineering activities considered necessary to achieve a trusted computer system. These tasks are strongly based on guidelines developed by the MITRE Information Security Center (with additional collaboration from the companies listed in section 4.2), on how to incorporate security requirements in the system specification and in the statement of work for systems in classes B1 through A1. Further refinements were made to these guidelines to include the two classes found in division C.

Notice that the majority (14 out of 16) of the SOW tasks shown in Fig. 3–1 are intended to satisfy the assurance requirement defined in the Orange Book. Assurance provides the basis for proving that the developed security mechanisms cannot be compromised in an operational environment. Consequently, it is appropriate that the focus of the majority of tasks in the SOW be geared towards achieving assurance. Other tasks shown in Fig. 3–1, such as developing the security concept of operations (CONOPS) and performing a clandestine vulnerability analysis (CVA), are not explicit requirements in the Orange Book. However, they have been included to reflect completeness and current engineering practices. Table 3–3 provides a brief descriptive summary of the engineering activities defined by the statement of work tasks presented in Fig. 3–1.

Figure 3–2 presents the generalized security engineering activity schedule derived from the SOW tasks shown in Fig. 3–1. The schedule was developed and refined to its present level of detail through numerous discussions with security engineering experts from across the country.

The statement of work tasks described above have been related to the major review milestones of a DOD–STD–2167 software acquisition, to develop the generalized class D through A1 (shown in Fig. 3–2 by the notation D:A1) security engineering activity schedule presented in Fig. 3–2. The activities shown on the schedule reflect tasks that should ideally be accomplished prior to and following the indicated major review milestones.

The following characteristics of the security engineering process are provided by this schedule:

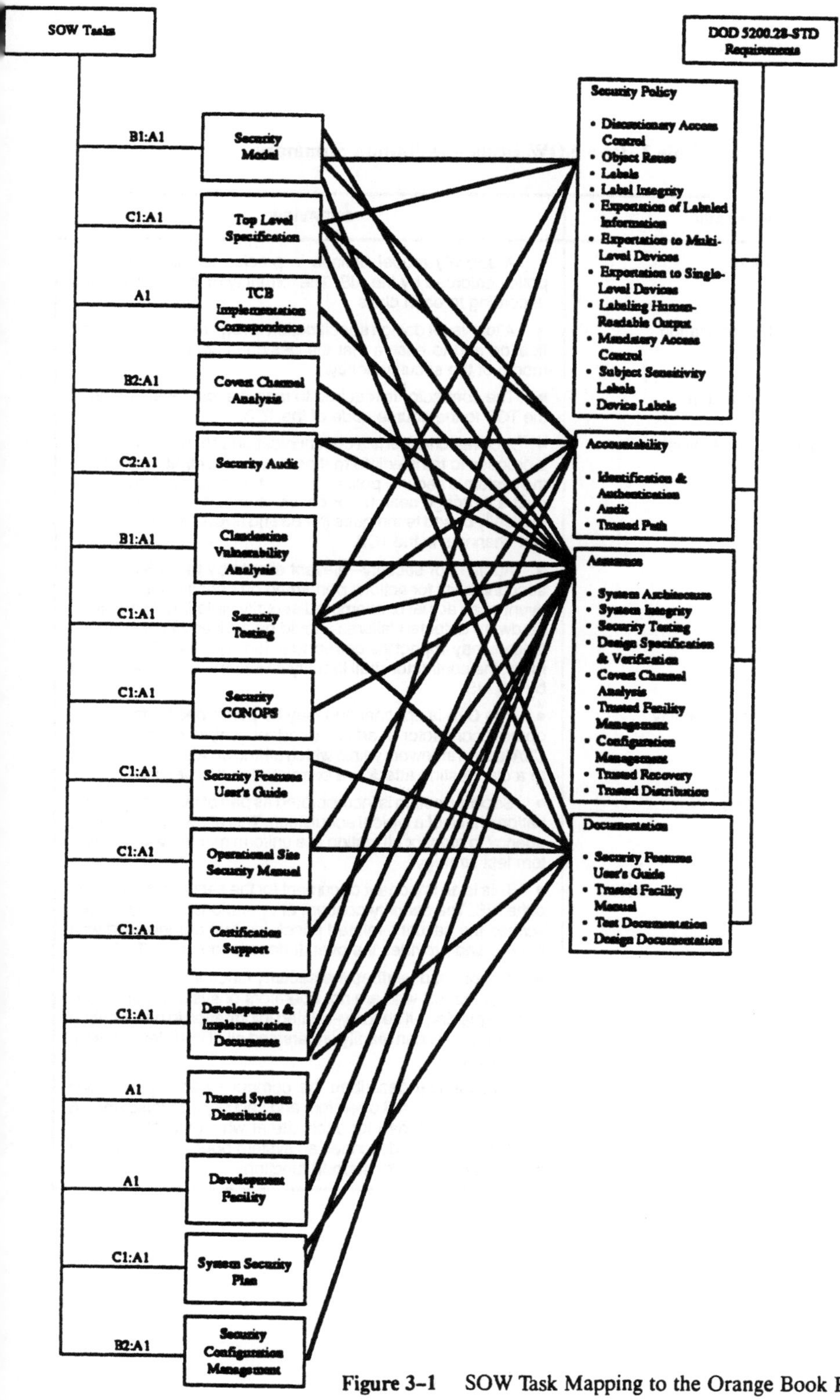

Figure 3–1 SOW Task Mapping to the Orange Book Requirements

Table 3-3 SOW Tasks - Definition Summary

SOW TASK	OVERVIEW
1. Security Model	• A security model defines in precise terms the security policy enforced by the AIS. The formality of the model varies according to each class.
2. Top-Level Specification	• A top-level design specification is required to assist certification and to ensure that the design is consistent with the model of the security policy.
3. TCB Implementation	• The contractor is required to map the top-level design of the TCB to the source code of the TCB.
4. Covert Channel Analysis	• A covert channel is a communications channel that allows a process to transfer information in a manner that may violate the system's security policy. The contractor is required to conduct a thorough search for covert channels and use mathematically based techniques (for B3 and above) that identify covert channels in the TCB.
5. Security Audit	• Auditing of security-relevant events by the AIS provides accountability for actions performed on behalf of a user in the event of an actual or attempted security violation. In addition, hardware detected failures should be audited because these failures may impact the correct functioning of the TCB. Furthermore, the contractor shall identify all information to be audited by the AIS.
6. Clandestine Vulnerability Analysis (CVA)	• The CVA is optional and may be performed by the development contractor or an independent contractor or both. The CVA is the framework within which all known AIS vulnerabilities to a clandestine attack are correlated and assessed.
7. Security Testing	• Security testing is incorporated as part of the test and evaluation phase of a typical acquisition. This allows the acquiring agency to monitor all testing in a uniform manner with the system test strategy.
8. Security Concept of Operations	• This is an important document for the security certification ofthe AIS. The security concept of operations (CONOPS) describes the security-related principles that are involved with the AIS and provides the overall description of the TCB.
9. Security Features User's Guide	• The guide describes the security functions performed by users of the AIS who are not operators or system administrators. By placing all security-related information in one location a potential user can readily understand all of the relevant security features.
10. Operational Site Security Manual	• This manual describes the computer security functions performed by the AIS operators and security administrators. A manual is developed for each site at which the AIS is to be used. The manual describes all system operator and system security administrator support functions.

Table 3–3 SOW Tasks – Definition Summary (Continued)

SOW TASK	OVERVIEW
11. Certification Support	• Contractor operational and technical support is required to assist government conducted certification efforts. The contractor is also responsible for maintenance throughout the certification process. The contractor often provides certification support through participation in the Security Working Group.
12. Development & Implementation Documents	• The contractor is requested to consolidate all security-related items in the type B and type C specifications.
13. Trusted System Distribution	• Trusted system distribution is necessary to insure that software, firmware, and/or hardware updates are not modified by unauthorized persons en route to their destinations. This is a requirement for A1 certification.
14. Development Facility	• This is optional and not a requirement in the Orange Book. Since continuous protection and trusted distribution are required for class A1 systems, some form of controlled environment for the development of the TCB should be required by the contractor.
15. System Security Plan	• The system security plan defines the contractor system security engineering program. The program identifies methods to be employed to identify security requirements, to synthesize and evaluate proposed solutions, and to contribute security inputs to the system development process.
16. Configuration Management	• Provides assurance that the security–relevant aspects of the system are protected from the introduction of malicious hardware and software during the developmental and operational life of the AIS.

- The acquisition milestone suggested for a specific security engineering statement of work task,
- Visibility into the security engineering tasks that become additional requirements as systems evolve from class D through class A1,
- A framework for cost estimating, and explicit identification of the specific elements of security engineering that cause the increase in cost as systems move toward the higher evaluation classes,
- A mechanism suitable for tailoring future schedules according to the specific security requirements of a program.

The generalized schedule shows that security–related activity is an essential part of the concept planning and definition phase of a program. Early planning for the impact of a program's security requirement(s) on the functional, architectural, and operational aspects of the system is crucial in achieving successful accreditation. In general, incorporating security features, a poster-

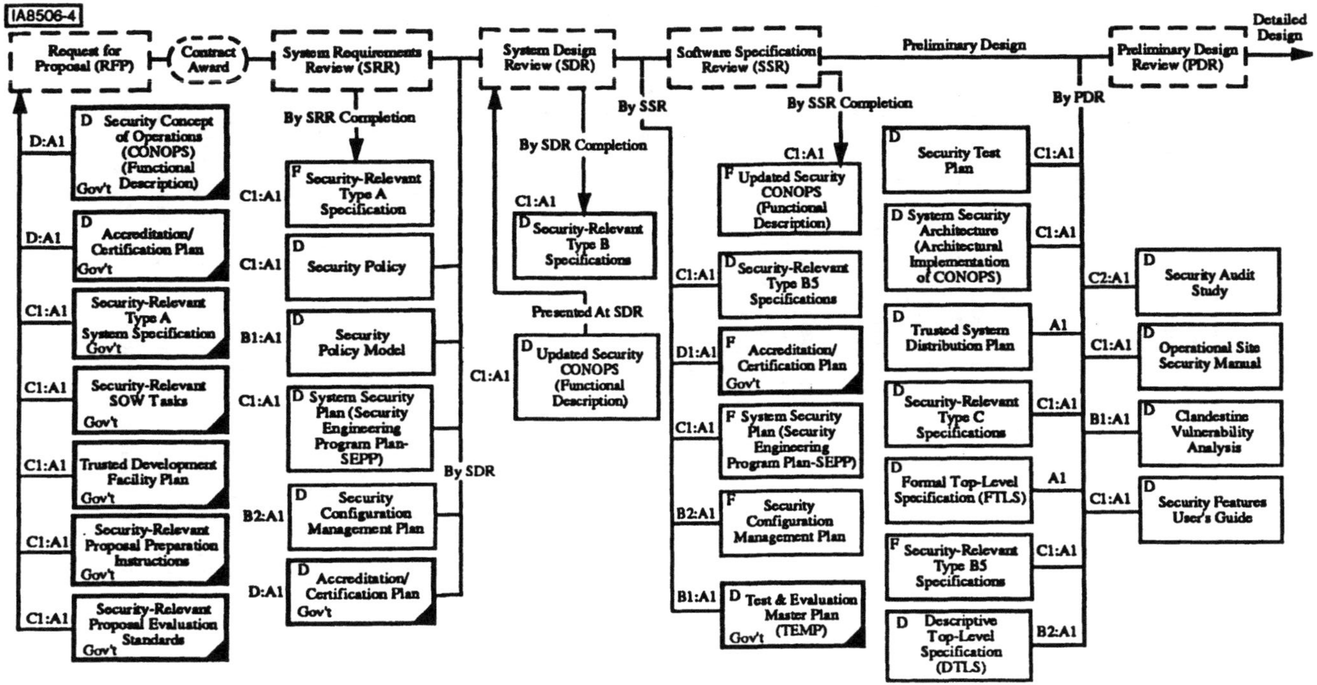

Figure 3–2 The Generalized Security Engineering Task Schedule

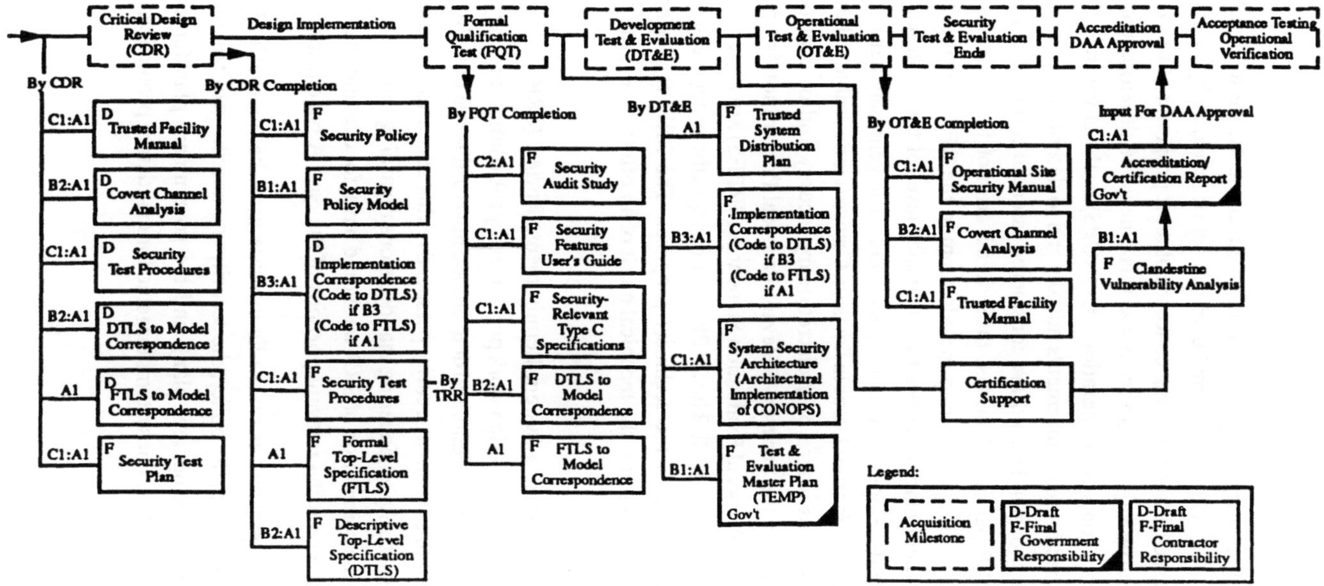

Figure 3–2 The Generalized Security Engineering Task Schedule (Continued)

iori, into an existing design is extremely difficult, especially for systems in the higher evaluation classes.

There are seven security–related items included in the Request For Proposal (RFP) package shown in Fig. 3–2. These items make up the set of products by evaluation class that thoroughly address the security engineering concerns of most programs at this early acquisition phase. The primary security engineering activity at this time is the development of a functional description of the security concept of operations (CONOPS) by the government. The CONOPS is needed to correctly specify the requirements initially and to determine the system vulnerabilities. The first four items shown in Fig. 3–2 are considered essential elements of the security–relevant portions of the RFP package. The last three items are not found in the Orange Book; however, it is good engineering practice to include them in the RFP package. Specifically, the trusted facility development plan describes the clearance level of the contractor development facility required for the development of secure software. It is generally needed for multilevel systems; however, we show it for the classes in Division C since it is desirable to have such a secure facility even in circumstances where only discretionary security and/or controlled access are required. The last two items, the Security–Relevant Proposal Preparation Instructions (SRPPI) and the Security–Relevant Proposal Evaluation Standards (SRPES) are either found together in the RFP or they are not included at all. The SRPP identifies specific system security issues that an offerer must explicitly address in the technical proposal. The SRPES is a source selection sensitive document. It describes how the technical merit of an offerer's security proposal should be judged.

Following contract award, a System Requirements Review (SRR) is held to review the system requirements analysis and design work performed by the contractor. The contractor must have evaluated the system requirements and determined how to allocate them into hardware and software configuration items. From a security perspective this allocation is critical since it is at this point in the system security engineering that decisions regarding the location of the trusted elements are made. By the completion of SRR we should have, for evaluation classes C1 through A1, a final version of the security–relevant portions of the type A system specification.

Security engineering activities markedly increase with the approach of System Design Review (SDR). Drafts of the security policy, the security engineering program plan (the contractor's security engineering development plan), and the government plan for certification and accreditation becomes available. For systems targeted for mandatory protection (division B) and higher, a draft of the security policy model defining in precise terms the security policy to be enforced by the AIS is also completed. The formality of this model, and, therefore, the effort required for its development, varies according to the specific class. Also available in draft form is the security configuration management plan for classes B2 and higher. Configuration control is required for the management of development and changes to the TCB to assure that it is protected from the insertion of malicious logic. Presented at SDR is a draft of the contractor's updated security CONOPS written at the functional description level. The contractor and the government must reach agreement on the security CONOPS early, since it may directly affect the degree of formalism required of security policy for the developing AIS. All security-related SDR action items must be later incorporated into the final version of this document. By the completion of SDR, drafts of the security-relevant type B specifications for classes C1 through A1 should be available.

With the approach of Software Specification Review (SSR), effort has been expended to complete three items, namely, the government accreditation/certification plan, the security engineering program plan, and the security configuration management plan. A draft of the security-relevant portions of the type B5 specification becomes available and reflects how the proposed software design will meet the security requirements of the AIS. Also in draft form is the government developed Test and Evaluation Master Plan (TEMP). A TEMP is written for systems targeting mandatory protection and higher for the purpose of specifying security test scenarios that are independent of the contractor's security test strategy. The TEMP evolves throughout the development of the AIS to incorporate changes in the system definition and does not become final until all action items following Formal Qualification Test (FQT) are resolved. Following the completion of SSR the updated security CONOPS becomes finalized.

As the developing AIS evolves towards Preliminary Design Review (PDR) the security engineering effort is focused on refining the software design specifications. Depending on the specific evaluation class as many as ten security engineering activities are taking place by the contractor. In particular, the type B5 specification is finalized and drafts of the security test plan, the system security architecture, the trusted system distribution plan, the security-relevant portions of the type C specification, the security audit study, the operational site security plan, the security features user's guide, the clandestine vulnerability analysis (CVA), and, for class A1 systems, the Formal Top-Level Specification (FTLS) become available. The FTLS is written in a formal mathematical language so that theorems showing the correspondence of the system specification to its formal requirements may be proven. For class A1 systems a draft FTLS is required by PDR. The FTLS is intended to assist with certification and to ensure that the design is consistent with the model of the security policy. The contractor must rigorously verify that the FTLS is consistent with the formal model of the security policy through formal verification techniques. This verification can require a large effort as will be shown in section 4. The CVA is not a requirement in the Orange Book; however, the analysis provides the framework within which all known AIS vulnerabilities to a clandestine attack are correlated and assessed.

With the AIS well into its detailed design phase, as many as four additional security engineering activities appear by the time Critical Design Review (CDR) is reached. The contractor's security test plan is finalized; drafts of the covert channel analysis (classes B2 and higher), the security test procedures, and the specification-to-model correspondence (class B2 and higher) become available. The effort required for the specification-to-model correspondence varies according to the specific target class. If the AIS is targeted for class B2 or B3 then this engineering activity is referred to as the Descriptive Top-Level Specification (DTLS)-to-model correspondence and it is typically found as part of the type C5 specification. The DTLS is written in a natural language (e.g., English), an informal program design notation, or a combination of the two. Furthermore, the DTLS might consist of the security-relevant portions of the B5 and C5 specifications. If the AIS is targeted for class A1 then this is referred to as the FTLS-to-model correspondence which can be mathematically formal. By CDR completion four engineering activities become finalized. They are the security policy, the security policy model, the security test procedures, and the FTLS (for class A1). Furthermore, if the AIS is targeted for class A1 then there is additional security engineering activity at this time to establish a draft implementation correspondence, which is a formal mapping of the FTLS to the TCB source code to assure the correct implementation of the security policy model.

By the completion of FQT up to four security engineering activities are finished. With all FQT action items resolved, the security audit study, the security features user's guide, the security-relevant portions of the type C specifications, and the model correspondence are now finalized as the evolving AIS proceeds into Development, Test and Evaluation (DT&E).

An important government activity that becomes finalized by DT&E is the test and evaluation master plan. Having been drafted by SSR, the TEMP has now been revised and modified to reflect any change in the contractor's security test plans and procedures that may have resulted from FQT action items. Furthermore, by DT&E the system security architecture (which is the architectural implementation of the CONOPS) has become finalized, and in the case of a class A1 AIS, the security engineering effort to verify the code–to–FTLS correspondence and to develop the final version of the trusted system distribution plan has been completed. If the AIS is targeted for class B2 or higher, then by the completion of Operational, Test and Evaluation (OT&E), final versions of the operational site security manual and the covert channel analysis become available. This analysis is a search for all covert means to violate the system's security policy through communications channels and it requires the use of techniques that are mathematically based (for B3 and higher) to identify covert channels within the TCB. Covert channel analysis requires considerable effort as will be shown in section 4. Certification support by the contractor continues throughout the system testing process upon completion of acceptance testing/operational verification. The Trusted Facility Manual (TFM) should also be completed if testing is being performed since it states how to bring up the system in a secure manner and maintain the system security during operations.

With the completion of Security, Test and Evaluation (ST&E) the AIS moves towards accreditation by the DAA. The DAA is the individual who officially authorizes the AIS to process sensitive information in its operational environment. The authorization is based upon a comprehensive security evaluation of the system's hardware, firmware, software, and other system–related procedural, administrative, physical, TEMPEST, personnel, and communications security controls. The government submits as input to the DAA its final accreditation/certification report which demonstrates that the AIS meets all the security requirements of the system. The CVA, which documents the system's security functional requirements and the residue clandestine vulnerabilities, is incorporated into the final accreditation/certification report.

4. IMPLEMENTATION OF THE SCHEDULES FOR COST ANALYSIS

Extension of the security engineering schedule, presented in section 3.1, as a framework for cost analysis is the theme of this section. The generalized task schedule described in section 3.1 organized the security engineering activities as a set of statement of work tasks and documents scheduled around the major milestones of a generalized system acquisition. Thus, we have a framework for developing bottom–up level–of–effort (LOE) engineering estimates of the resources associated with each activity in the schedule. Furthermore, the design of the schedule allows for tailoring the activities to the unique security requirements and statement of work tasks associated with the development of a particular project.

This section presents a mapping of the security engineering statement of work tasks shown in Fig. 3–1 to their respective cost elements in MIL–STD–881A. The results of a cost survey, which was formulated around the schedule presented in section 3.1, are provided in this section. Furthermore, the data obtained from the survey provide a view into some of the differences in the engineering effort between implementing class B1 versus class A1 security requirements.

4.1 Mapping Security Engineering Activities to MIL–STD–881A

Section 3 described the link between the security requirements in DOD 5200.28–STD to the schedule of security engineering statement of work tasks and document deliverables as they relate to a DOD–STD–2167 acquisition. These connections support technical program planning and the scheduling of security–relevant events around the major development milestones of a program. To support the implementation of this schedule to cost analysis, Fig. 4–1 connects the security engineering tasks, shown in Fig. 3–1, to their respective MIL–STD–881A cost elements.

The security engineering tasks identified in this study are associated in Fig. 4–1 with four MIL–STD–881A cost elements: system engineering, system test and evaluation, data, and operational site activation. Most of the SOW tasks involve some type of engineering design and/or analysis activity, and are, therefore, appropriately linked with the system engineering element as defined in MIL–STD–881A. Notice from this mapping that the data cost element appears not only by itself, but also with the system engineering and system test and evaluation cost elements. In general, when the security engineering tasks of the type identified in this study are implemented on a project, they ultimately yield a series of products that are required to be formally documented. Documentation is a vitally important dimension of security engineering. The evidence presented in the final documentation regarding the delivered system's security is among the major criteria used when independent evaluators assess, during the certification process, the degree of trust that can be expected in the operational environment.

Security testing falls under the general system test and evaluation cost element, and it encompasses many important activities. Among these are Security Formal Qualification Test (SFQT), Security Qualification Test and Evaluation (SQT&E), and Security Qualification Op-

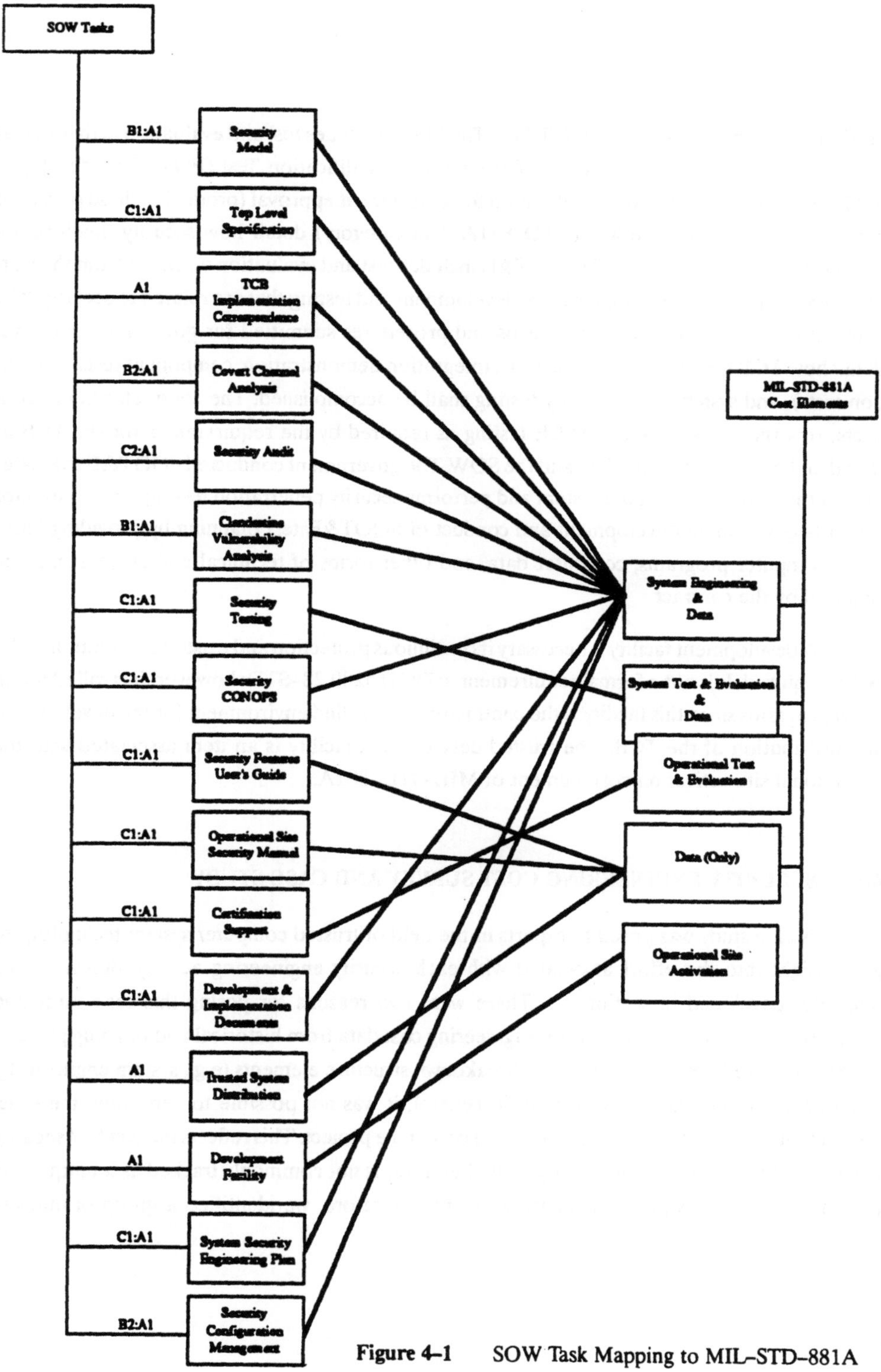

Figure 4–1 SOW Task Mapping to MIL-STD-881A

erational Test and Evaluation (SQOT&E). The SFQT includes test and evaluation activities that meet the security-relevant portions of the Formal Qualification Test (FQT). The developed SFQT test plans and procedures submitted for government approval (producing these plans and procedures falls under the MIL-STD-881A data category) detail how security qualification testing shall be accomplished. The SQT&E includes test and evaluation activities from the start of system design through completion of development and testing that are related to security. The contractor-developed SQT&E test plans and procedures submitted for government approval detail how TCB qualification testing (e.g., integration demonstration, component tests, integration tests), and system level security testing shall be accomplished. The contractor plans, conducts, reports, and supports SQT&E testing as required by the requirements for the TCB as stated in the system specification, and the SOW. The government conducted SQOT&E evaluates the operational security of the system and performs security penetration testing. The contractor supports government development and conduct of SQOT&E test planning by providing hardware, computer programs, computer data, and other forms of technical assistance as may be required by the contract.

A development facility is necessary if continuous protection and trusted distribution of the AIS is required. It is not a formal requirement in DOD 5200.28-STD; however, it is called for on many programs since this facility is the contractor's controlled environment for the development and distribution of the TCB. The trusted development facility is an item associated with the operational site activation cost element of MIL-STD-881A.

4.2 SECURITY ENGINEERING COST SURVEY AND CASE STUDY

A case study was posed to experts in the field of trusted computer system technology to gain insight into the effort associated with each security engineering activity shown on the schedules developed in section 3.1. There were two reasons for posing this case study and conducting the survey. First, security engineering cost data from historical and ongoing projects were so scattered throughout the work breakdown structure elements (e.g., system engineering, system test and evaluation, data) that, in general, it was not possible to determine the exact impact that security requirements had on the cost of the project. This reflects the fact that security engineering is a system-wide activity, and, therefore, is not commonly tracked as a unique line item in a program's work breakdown structure. Therefore, we identified a group of industry experts with experience in the design, development, test, and verification of trusted computer systems to participate in a cost case study. Second, we were interested in assessing the differences in effort to build a trusted computer system with class B1 features versus class A1, which offers the highest level of protection and a design that is formally verified through the strictest set of techniques currently employed by the industry. This difference, viewed at the security engineering task level shown in the activity schedules, could not be seen from only the historical program cost data which we had access to collect. Participation in the LOE survey included the following organizations:

- The MITRE Corporation
- TRW
- Honeywell Federal Systems, Inc.
- Trusted Information Systems, Inc.
- Intermetrics, Inc.

Participants were given a package of materials around which the LOE analysis was based. The package included: a technical problem statement for a representative B1 and A1 system, a list of security engineering activities for classes B1 and A1 excerpted from the schedules shown in Fig. 3–2, a detailed set of SOW task descriptions for each activity on the class B1 and class A1 schedule, a data collection form, and the class B1 and class A1 security engineering activity schedules derived from Fig. 3–2.

We described a hypothetical system that was representative of a baseline security engineering problem, that is, the system security requirements did not extend beyond DOD 5200.28–STD, and the system security perimeter consisted of a single host processor which users accessed through terminals and where all data processing was to be performed.

Figure 4–2 shows a high level view of the system hardware architecture around which the cost case study was based. The hardware architecture consists of a host processor which users access through terminals. All processing is done on the host processor. The system receives messages from and sends messages to a multilevel network through a communications front–end processor which is part of that network.

The security features of this system are required to prevent the compromise of classified information, minimize overclassification of information, prevent unauthorized alteration of

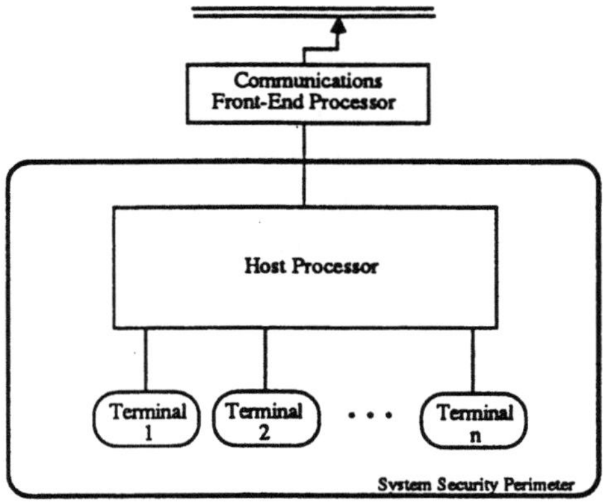

Figure 4–2 System Hardware Architecture – Case Study

information, and prevent denial of access to information or resources to which a subject is entitled. In the case of B1, this system is required to operate in multilevel mode (all users are cleared for at least confidential) and will process SECRET data. In the case of A1, this system will operate in multilevel mode (there are cleared and uncleared users) and will process SECRET data. In both the B1 and A1 cases, it was assumed that no commercial general purpose secure operating system was available for this system.

The specific requirements for the B1 and A1 system are listed in Table 3–2 and they reflect the minimum requirements set called for in the TCSEC. Notice that there are eight additional requirements for the class A1 system above the 19 shown in Table 3–2 for class B1. These are primarily due to the stricter set of assurance requirements called for in class A1, which include such items as: the covert channel analysis, the management of a trusted development facility, adherence to a strict set of configuration controls (to maintain changes to the formal security policy model and the descriptive and formal top level specifications), in addition to requirements for trusted distribution and trusted recovery following an unanticipated system failure or shut-down. The functional requirements at class A1 are more demanding than those at class B1. This is reflected in the number and nature of shalls in the system A–level specification rather than in the SOW tasks.

Table 4–1 presents the list of class B1 and class A1 security engineering activities upon which this survey was conducted. These activities were derived from the schedules developed in section 3.1. Following the task numbering scheme in Table 4–1, effort estimates for the set of class B1 tasks and effort estimates for the set of class A1 tasks were collected from each survey participant.

The survey participants were requested to estimate the *total resource* in staff–months required to perform all aspects of the security engineering activities defined in Table 4–1, and to include all management reviews leading to the completed tasks and/or final documents. The effort estimated for any security engineering activity that resulted in a document or plan that is common to an acquisition (e.g., configuration management plans, test plans and procedures) was to reflect the effort above what would normally have been expended. Furthermore, the effort estimated for developing the security test plans and procedures was to include the effort to implement those procedures as part of the normal system test and evaluation phase of a program. In general, the estimates were to reflect the marginal increase in effort to perform the activities listed in Table 4–1 over and above what would normally have been anticipated without security requirements. Estimates were to be based on the assumption that all prior security engineering activities that may relate to the start of a particular task had been successfully completed.

For each task a range was given which reflected a pessimistic, most likely, and optimistic effort estimate. We were not seeking the absolute maximum or absolute minimum estimates but rather a feasible range that covered the estimator's experience. The optimistic estimate reflected a low but possible number of resources that an estimator felt could successfully complete the task. The most likely estimate reflected a level of resources that an estimator expected to see in most cases. The pessimistic estimate reflected a very conservative number of resources that an estimator felt could successfully complete the task.

Estimators were allowed to confer with other members in their organization who were technically expert in a particular activity or set of activities for their input into this LOE survey. For each activity the estimators indicated whether its effort estimate was based on a subjective assessment or from the historical experience of a known program. A subjective assessment was cited in circumstances where the estimate reflected a judgement based on limited experience with that particular task. Historical experience was cited in circumstances where the estimate was based on a known or ongoing program.

Table 4–1 Security Engineering Activity List for Classes B1 and A1

SECURITY ENGINEERING ACTIVITY (GOVERNMENT, CONTRACTOR)		EVALUATION CLASS B1	A1
Number	Task		
1.	Incorporating the Security-Relevant tasks into the Statement of Work (SOW). (Government)	*	*
2.	Developing the Trusted Development Facility Plan. (Government)	*	*
3.	Formulating the Security-Relevant Proposal Preparation Instructions. (Government)	*	*
4.	Establishing the Security Relevant Proposal Evaluation Standards. (Government)	*	*
5.	Developing the Security-Relevant Portions of the Type A System Specification for inclusion in the Request for Proposal Package. (Government)	*	*
6.	Establishing the Accreditation/Certification Plan. (Government)	*	*
7.	Developing the System Security Plan - the Security Engineering Program Plan (SEPP). (Contractor)	*	*
8.	Developing the Security Configuration Management Plan. (Contractor)	.	*
9.	Writing the Updated Security CONOPS (Functional Description). (Contractor)	*	*
10.	Incorporating the Security-Relevant features into the Type B specifications. (Contractor)	*	*
11.	Establishing the Security Test Plan. (Contractor)	*	*
12.	Formulating the Security Policy. (Contractor)	*	*
13.	Developing the Security Policy Model. (Contractor)	*	*
14.	Establishing the Security Test Procedures. (Contractor)	*	*
15.	Writing the Descriptive and Formal Top Level Specification (DTLS and (FTLS). (Contractor)		*
16.	Performing the Security Audit Study. (Contractor)	*	*
17.	Developing the Security Feature User's Guide. (Contractor)	*	*
18.	Incorporating the Security-Relevant features into the Type C specifications. (Contractor)	*	*
19.	Performing the FTLS-to-Model Correspondence. (Contractor)		*
20.	Establishing the Trusted System Distribution Plan. (Contractor)		*
21.	Implementation (Code-to-FTLS) Correspondence. (Contractor)		*
22.	Developing the System Security Architecture (Architectural Implementation of CONOPS). (Contractor)	*	*
23.	Formulating the Test and Evaluation Master Plan (TEMP). (Government)	*	*
24.	Writing the Operational Site Security Manual. (Contractor)	*	*
25.	Performing the Covert Channel Analysis. (Contractor)		*

Table 4–1 **Security Engineering Activity List for Classes B1 and A1 (Continued)**

SECURITY ENGINEERING ACTIVITY (GOVERNMENT, CONTRACTOR)	EVALUATION CLASS B1	A1
Number Task		
26. Writing the Trusted Facility Manual. (Contractor)	*	*
27. Writing the Accreditation/Certification Report for submittal to the DAA. (Government)	*	*
28. Performing the Clandestine Vulnerability Analysis (CVA). (Contractor)	*	*
29. Contractor support furing Certification process. (Contractor)	*	*
Number of		
Government Activities:	8	8
Contractor Activities:	15	22[1]
Overall Activity Total	23	30

[1]In class A1 task number 15 includes two efforts, namely, writing the DTLS and the FTLS.

4.3 SUMMARY FINDINGS

A summary of the LOE survey results is shown in Figs. 4–3a and 4–3b. These figures present the distribution of the average efforts (the average of the optimistic, most likely, and pessimistic) estimated for each task by the survey respondents. The number of points shown for a given task is equal to the number of responses received for that task. These points reflect the average of the optimistic, most likely, and pessimistic estimates. No effort estimates are shown in Fig. 4–3a for the security engineering tasks 8, 15, 19, 20, 21, and 25 since these tasks pertain only to class A1, in addition, the class B1 effort estimated for task 11 includes task 14.

The sample mean and variance of the average effort estimates presented in Figs. 4–3a and 4–3b were determined for each task, and then summed across all the class B1 and class A1 tasks to yield an overall effort mean and standard deviation for class B1 and class A1. Table 4–2 presents this summary.

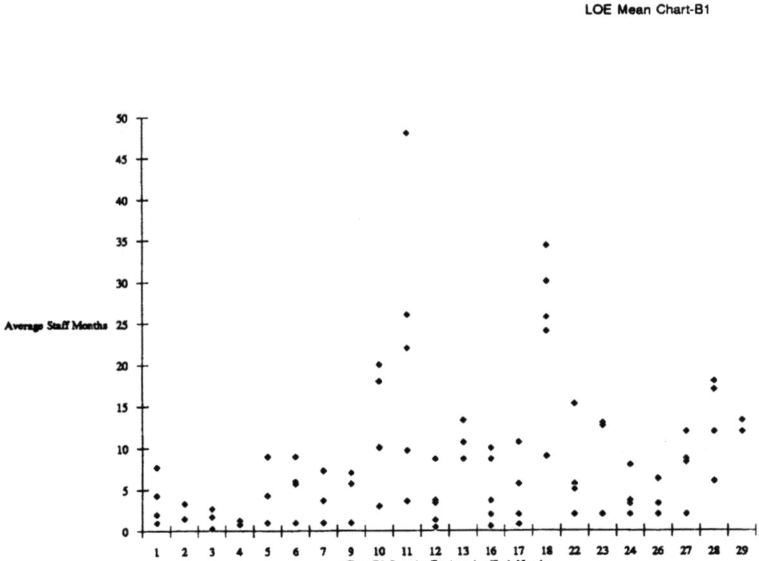

Figure 4–3a Distribution of Average Effort Estimates for Class B1

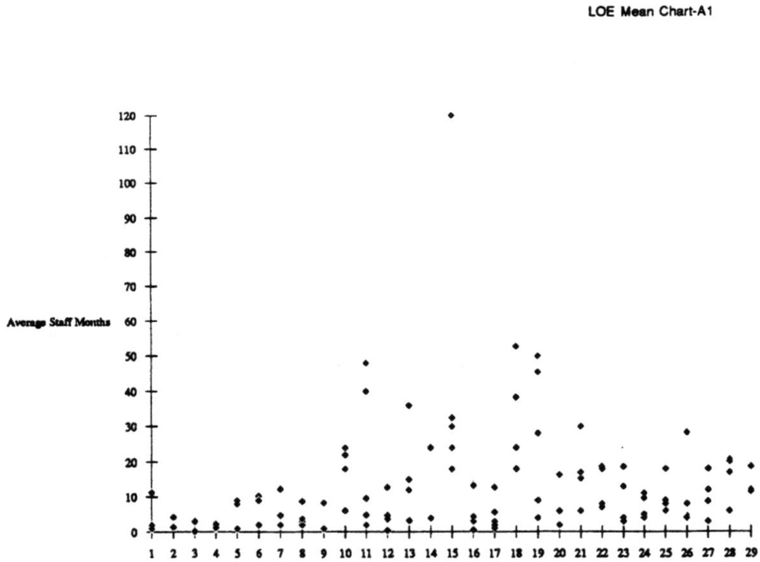

Figure 4–3b Distribution of Average Effort Estimates for Class A1

Table 4–2 **Survey Data Summary (Staff Months)**

SECURITY CLASS	GOVERNMENT & CONTRACTOR TASK	CONTRACTOR TASKS ONLY
B1		
Mean	166	132
Standard Deviation	26	24
A1		
Mean	348	303
Standard Deviation	64	62

Dependent upon a respondent's project experiences, larger estimated efforts tended to be associated with programs where the security requirements were part of a larger set of system wide requirements. In such programs, the security requirements had to be integrated with the other functional and performance requirements of the system. Furthermore, it is not uncommon for these types of programs to require security functionality that is in addition to the minimum requirements set defined in the TCSEC. For instance, a system might require that hardware, firmware and/or software features be developed to periodically validate the correct operation of the TCB, or it may be a requirement that the TCB software components be encapsulated from the non–TCB software. In these circumstances, security engineering would have to spend additional resources working on the design the system's TCB to meet these types of system–unique functional requirements. Therefore, tasks such as developing the security policy model, incorporating the security–relevant features of the TCB into the type B and type C specifications, and security support to system testing are among the security engineering activities that would reflect the effort increase. Thus, it is the extent to which the sophistication of a system's security requirements pushes the state of trusted computer system technology that has the singular effect on the magnitude of the security engineering effort invested. Respondents who experienced smaller efforts for the security engineering tasks identified for this case study tended to be associated with projects whose mission involved building a secure operating system and where the system security requirements were almost exclusively derived from a specific TCSEC class (e.g., a contract to develop a class C2 or class A1 general purpose secure operating system).

From Fig. 4–3a, it can be seen that for class B1, tasks 10, 11, 18, and 28 at least one respondent estimated an average effort over 20 staff–months. We will refer to these as the class

B1 cost drivers. It's not surprising to find, when these task numbers are matched to their definitions in Table 4–1, that the B1 cost drivers are the efforts to incorporate security-relevant features into the type B and type C design specifications, establishing the security test plans and procedures, and performing the clandestine vulnerability analysis. These are the engineering activities integral to design decisions on how security requirements are incorporated into the final system architecture and in determining and structuring the appropriate security test philosophy.

The class A1 cost drivers 13, 14, 15, 19, 21, and 26, shown in Fig. 4–3b, are in addition to those found for class B1. These tasks represent the effort to develop the security policy model, establish the security test procedures, write the FTLS, perform the FTLS-to-Model Correspondence, perform the Code-to-FTLS Correspondence, and write the Trusted Facility Manual (TFM). Notice that the effort for task 15 ranged up to 120 staff months. One of our respondents cited an actual experience on a secure operating system development that was targeted for A1 certification and placement on the EPL, of 5 equivalent full time staff working for two years (120 staff months). This was a labor intensive effort to develop all parts of the specification, prove the correctness of the security theorems with respect to the model, and iterate through all review changes until the final document for the formal top-level specification was completed.

These tasks have been deemed drivers only in the sense that the average of their ranges has been estimated by at least one member of our industry survey group to exceed an effort threshold of 20 staff months. The purpose of identifying these tasks is to show which security engineering tasks have the potential to drive larger efforts relative to the other tasks in Table 4–1. Therefore, these are the tasks that we recommend deserve careful treatment when assessing their efforts in a practical cost estimating situation.

The estimates summarized Figs. 4–3a and 4–3b demonstrate that the security engineering activity schedule presented in section 3.1 provide a legitimate framework for deriving engineering estimates of the effort to build trusted computer systems. It is recommended that the security cost and technical database established by this research continue to evolve as systems presently implementing Orange Book requirements near certification. The more data collected on a particular security engineering task, the more likely it is that the effort variance associated with that task will be lessened from what some of the ranges in this study have suggested.

In practice, viewing the effort data against a program's acquisition schedule, requirements drivers, and technology constraints provides a basis for determining the estimate considered most appropriate to the security engineering task. Analysts should not feel constrained to remain within the effort ranges provided by this study if sound technical justification exists. It is important to emphasize that the data collection aspects of this research is in its infancy. Thus, its initial value at this point in time is in providing the estimator with a reference base for developing security engineering level-of-effort estimates where none had previously existed. Engineering judgement will always be integral when assessing the similarity and technical complexity of each security engineering task for the system being costed against the data sets made available by this research.

In addition to the system-level engineering efforts identified above, it is also important to incorporate the impact that a system's security requirements may have on the cost of the hardware and, in particular, on the software development resources. Frequently, it is the mission requirements of a system (e.g., performance, fault tolerance) rather than its specific security requirements, that dictate the hardware architecture. However, software development can be directly impacted by security requirements in a number of ways. For instance, where the trusted elements will reside in the system (e.g., in the workstations, in the LANs, in the host processor, or some combination of these) can have major impacts on the final software and security architecture, the software design approach, and the testing philosophy.

Software development can also be impacted by whether the individual application programs will be benign or trusted. Trusted software is responsible for enforcing security, and therefore, the security of the system ultimately relies on its flawless and highly reliable operation. The need to develop trusted applications software can, therefore, impact the overall amount of code to be developed and tested. The software development language is also important to consider from a security perspective. For instance, security requirements can force a restricted use of certain Ada language features for the trusted portions of the system. These are just a few of the areas where security requirements can impact the development and testing of software. The extent that software development is impacted by security must be examined on a case basis, that is, as a function of the system's unique set of mission performance and security requirements. Furthermore, this research has not seen evidence to suggest that estimating the effort to develop software security functions using a software resource model (e.g., COCOMO [3]), should be different from current practices. This presumes that the model parameters are appropriately gauged to reflect any additional impacts that security specific requirements may have, such as increased levels of reliability testing, on the software development.

4.4 COST METHODOLOGY GUIDELINES

This section provides a general set of guidelines for the analyst implementing the results of this research to estimate the cost of building systems with DOD 5200.28-STD security requirements. The results from this research study consist of two parts: the general set of security engineering activity schedules developed for each TCSEC evaluation class, and the effort data collected for the tasks shown on these schedules. The framework provided by the schedule and the effort data is suitable for constructing LOE engineering estimates of the cost to build trusted computer systems, rather than predicting the effort mathematically from a set of input parameters.

Table 4-3 presents a set of guidelines which should be considered when costing the DOD 5200.28-STD security requirements of a system. They have been developed to assist practicing analysts in interpreting and applying the results of this research to the cost estimating environment. Finally, it is important to stress that estimating the cost to build trusted computer systems involves the collective technical expertise of the cost analyst and the system, security, and software engineers on the program. This is evident upon reviewing the guidelines presented in Table 4-3.

5. RECOMMENDATIONS FOR FUTURE RESEARCH

The research presented in this paper was conducted for the purpose of achieving two goals. First, we sought to develop a technique that would identify the areas where security requirements have a cost impact across the system. Second, determine what is driving these impacted areas and make some estimates on the magnitude of the cost. The cost estimation methodology presented in this paper is structured around the security engineering process involved with building a trusted computer system. Directly linking the cost estimation methodology to the specific elements of the security engineering process has provided a vehicle for

Table 4–3 Implementing the Estimation Approach – Guidelines for Costing

STEP	TASK	GENERAL COMMENTS
1	Determine the TCSEC evaluation class and the intended security operational mode required of the system.	This determination comes from the security engineering team.
2	Determine from the program's SOW the set of security engineering tasks required, and map these to the appropriate class (C1, C2, B1, B2, B3, or A1) schedule developed in section 3.1. Determine from this mapping the security engineering tasks that are either missing or are in addition to those already existing in the schedule.	Assumes that the program has sufficiently evolved in its acquisition planning to have a SOW and a draft system level specification.
3	Determine where to locate the additional security engineering tasks, discovered in step 2, to the major review milestones on the schedule. Resolve why there are security engineering tasks missing from the program's SOW when compared to the generic schedule developed in section 3.1.	Completing step 3 provides the project leader with a technical planning tool that identifies all the security engineering activities that must be scheduled with the other engineering activities in the system.
	Coordinate schedule with appropriate project technical staff and construct the final security engineering schedule tailored to the system.	
4	Perform an LOE estimate of the security engineering effort for the tasks on the final schedule constructed in step 3.	The effort data collected for this research can serve as a reference base for developing the LOE estimate. Engineering judgement is needed to evaluate the similarity and technical complexity of each security engineering task for the system being costed, against a similar set of security engineering tasks for which this research has made effort data available.
5	Obtain the impact of the system's security requirements on the size of the software to be developed. Determine whether any hardware development is required.	It is important to incorporate into the prime mission product costs, any effort that system security requirements may have on the cost of the hardware and software to be developed.Presently, this research has not seen evidence to suggest that estimating the effort to develop a software security function should be different from current methods.
6.	Map the LOE estimates for each security engineering task into the cost elements of MIL-STD-881A. Guidelines are provided in Fig. 3-1.	This mapping is important for correctly spreading the cost of security into the appropriate cost elements of MIL-STD-881A. This supports phase funding and assessing cost uncertainty, which are typically done at the element level.

measuring cost sensitivity to engineering trade-offs and to the often unique character of each system's security requirements.

The principal product of this research is a framework of the security engineering process that supports the technical program planning and the costing of systems implementing DOD 5200.28-STD security requirements. Prior to this investigation, we did not have the level of insight presented herein into how the cost and technical dimensions of the security engineering process are interleaved with the overall system engineering activities of a developing trusted computer system.

This research is the culmination of a five month study whose primary focus was to develop an estimating framework for costing the security requirements of the type found in the TCSEC. It was natural to begin here since the TCSEC is still the only official standard for security engineering in existence today. The Trusted Network Interpretation (TNI) [4] interprets and extends the concepts contained in the TCSEC to computer networks. Although the TNI has been recently issued, the National Computer Security Center (NCSC) will not consider it final until at least one year of vendor and evaluator experiences are incorporated in a planned revision to the document. The NCSC is also developing similar interpretations for data base management systems. These interpretations will be known as the Trusted DBMS Interpretations (TDI). Therefore, a natural extension of this study would be to construct security engineering task schedules and effort data bases for the TNI and TDI documents.

Finally, it is recommended that the following areas be considered in future cost research studies: analyzing an expanded set of programs either completed or nearly completed from a cost and technical perspective, that also includes programs targeting for system high mode (the most common in Air Force systems); obtain data from security product vendors to determine line of code estimates for a variety of types of trusted software (e.g., kernels, trusted applications code) and cost data associated with trusted hardware developments; examining the cost associated with developing trusted applications on commercial off-the-shelf (COTS) TCBs; examining the costs of modifying COTS TCBs to meet specific mission requirements; and analyzing the costs of modifying and augmenting COTS TCBs to meet higher criteria class requirements.

REFERENCES

1. "Department of Defense Standard: Department of Defense Trusted Computer System Evaluation Criteria," DOD 5200.28–STD, Department of Defense, Washington, DC, December 1985. (Revised March 1988).

2. "Military Standard Work Breakdown Structures For Defense Materiel Items," MIL–STD–881A, 25 April 1975.

3. Boehm, B. W., *Software Engineering Economics*, Prentice–Hall, 1981.

4. "Trusted Network Interpretation of the Trusted Computer Security Evaluation Criteria," NCSC–TG–005, Version 1, National Computer Security Center, Fort Meade, MD, 31 July 1987.

ACKNOWLEDGEMENTS

The author gratefully acknowledges the support of M. Daya, D. J. Bodeau, F. N. Chase, W. R. Gerhardt, and P. S. Tasker of The MITRE Corporation; R. T. Jordan of Intermetrics, Inc., C. Bonneau of Honeywell–Bull, Inc., E. Anderson of TRW, and T. C. V. Benzel of Trusted Information Systems. All of these individuals made significant technical contributions to this research study.

THE COST EFFECTIVENESS OF INTEGRATED AVIONICS ARCHITECTURES

by

Bruce Harmon
Institute For Defense Analyses

Steve Shyman
Boeing Aerospace and Electronics

ABSTRACT

The analyses presented in this paper examine the benefits and costs associated with integrated avionics architectures proposed for the next generation of tactical aircraft (as exemplified by the Advanced Tactical Fighter (ATF)) when compared to more traditional federated architectures. In an integrated avionics suite the system's processing capability is consolidated in one or two central processors; in a federated suite the processing capability is distributed throughout the system in black boxes associated with each avionics subsystem. We quantify benefits and costs in terms of dollars. This is possible when we compare integrated and federated systems that have similar functionality, i.e., where mission performance is comparable. The benefits and costs of integrated avionics are quantifiable as decreased hardware life-cycle costs and increased software life-cycle costs.

1. BENEFITS OF INTEGRATED AVIONICS

In this section we estimate the benefits of integrated avionics architectures. Benefits are quantified as decreases in the life-cycle costs of a hypothetical ATF-generation fighter associated with going from a baseline federated architecture to an integrated architecture, holding mission capabilities constant. We also enumerate non-pecuniary benefits where the attributes of integrated systems can be exploited to attain functionality not available in a federated system.

The cost savings are derived from avionics weight and volume savings inherent in an integrated avionics architecture when compared to a federated architecture. Cost savings are not a result of decreases in avionics hardware costs, but in decreases in airframe and engine costs resulting from decreases in overall aircraft weight.

We focus on the digital portions of the avionics suite. We assumed that the weight and cost of the analog "front-end" components of the hypothetical system would remain the same regardless of whether the processing associated with the functions of the analog components is done in one or two large integrated processors or is accomplished in a large number of smaller processors dedicated to each type of function (i.e., radar, electronic warfare, communications). When we compare processing architectures we also project the same level of technology for each type; it would be unfair to compare a federated system based on current technology (technology associated with existing federated systems) to an integrated system based on the advanced technologies available to next-generation systems.

We also assumed that the weight savings occur in the low-value (low cost-per-pound) portions of the avionics hardware such as the backplanes, communication ports, power supplies, and enclosures that form the "overhead" for each package of processor boards or modules. When processor functionality is consolidated into one or two large packages (black boxes) weight savings occur as the larger amounts of processing capability are spread over these overheads. Because of this we do not translate the weight savings into avionics hardware cost savings. For processor black boxes we expect decreasing returns to scale with respect to weight, while assuming fixed hardware cost.

In estimating the benefits of integrated avionics we proceeded in four steps.

1. We used linear regression analysis to estimate the relationship between processor capacity, vintage of technology, and processor weight. We expected processor weight to increase at a decreasing rate as processor capability increases, and to decrease with time, holding capacity constant.
2. Given this relationship, we estimated the processor weights for the following notional systems:
   a. A baseline federated system having the same major avionics functions and technology level expected in an ATF-generation fighter. This system would have individual processors dedicated to each function, where total memory capacity of the processors and the distribution of that capacity across functions would be consistent with that expected in an integrated ATF-generation system.

b. An integrated system with attributes similar to those expected for an ATF-generation fighter, including avionics functions, processor capacity, and number of integrated processors.

3. We then derived differences in overall aircraft weight and corresponding increases in engine thrust resulting from the difference in estimated avionics weights between federated and integrated architectures based on engineering rules of thumb, holding air vehicle performance constant.

4. Finally, we generated cost estimates for development, production, and operating and support (O&S) costs based on aircraft and engine parameters for both system architectures. The differences in these values represent the savings in life-cycle cost associated with integrated avionics.

Data used in the statistical analysis of processor weights included characteristics of processors used in fighter and attack aircraft ranging in age from the F-14A to the F-15E. These data are shown in Table 1-1. Processor characteristics of most interest include processor weight, processor capacity, and full-scale development (FSD) start year. In general, FSD start date is the same as the start date for development of the processor, except in cases where we had specific schedule data to the contrary.

Table 1-1 Processor Data Used in Regression Analysis

APPLICATION	DESCRIPTION	FSD START TEAR	WEIGHT (POUNDS)	CAPACITY (KILOBYTES)
F-14A	Fire-Control Computer	1970	166.0	128
F-15A	Central Computer	1972	39.6	64
F-15A	Radar Data Processor	1972	40.1	32
F-16A	Fire-Control Computer	1975	28.5	128
F/A-18A	Mission Computer	1976	41.0	256
F/A-18A	Air Data Computer	1976	17.4	12
F/A-18A	Radar Data Processor	1976	42.9	128
F/A-18A	Radar Signal Processor	1976	51.5	384
F/A-18A	Flight-Control Computer	1976	43.0	160
AV-8B	Mission Computer	1976	43.5	256
AV-8B	Air Data Computer	1979	15.9	24
AV-8B	Stores Management Processor	1979	30.4	104
AV-8B	Radar Warning Receiver Processor	1979	26.1	136
AV-8B	Armament Control Panel	1979	9.2	20
AV-8B	Aircraft Stabilization Computer	1979	15.1	48
AV-8B	Communication, Navigation, Identification Data Computer	1979	13.5	24
F-15E	Radar Data Processor	1983	57.0	2,048
F-15E	Radar Signal Processor	1983	59.8	2,048
F-15E	Central Computer	1984	55.0	3,000
F-16C	General Avionics Computer	1986	21.0	1,024

In estimating the multiplicative relationship between processor weight, processor capacity, and time, we estimated the coefficients for this equation by first transforming the equations to a log–log form and then applying standard linear regression techniques. When this is done, the constant term of the equation becomes biased when it is transformed from the log–log form back to a multiplicative form. We adjusted for this bias by adding one–half of the regression mean square error to the constant term of the log–log equation. This adjustment is an approximate correction for the bias introduced by the log–log estimation methodology. The resulting equation and measures of statistical significance and model fit are presented below:

$$\text{Weight} = 28.2\,(\text{Capacity})^{0.37}\,(\text{FSD Start Year}-1969)^{-0.81}$$

$$\qquad\qquad\qquad(0.01)\qquad\qquad\qquad\qquad(0.01)$$

$N = 20 \qquad R^2 = 0.91 \qquad \text{Adjusted } R^2 = 0.89 \qquad SEE = 0.22$

Significance levels are in parentheses below the parameter estimates. N is the number of observations and SEE is the standard error of the estimate. The regression results show all parameter estimates significant at the .01 level. The parameter estimate on the capacity variable is as expected, indicating that processor weight increases at a decreasing rate with capacity. The parameter estimate on the time variable is also consistent with expectations, indicating decreasing processor weights over time. Figure 1–1 plots processor weight predicted by the above equation against program actuals. Values above the 45–degree line are overestimated by the model, while values below that line are underestimated. The large outlying data point is the F–14A fire–control computer. Although it is very different in weight from the rest of the data base, regression diagnostics show that it is not an overly influential data point.

Figure 1–2 presents contour lines derived from the regression equation showing the relationship between processor capacity, development start year, and processor weight.

Once we obtained the estimating relationship for processor weight, the next step was to apply it to our hypothetical processor architectures. A notional ATF–type system provides a common baseline for overall processing capabilities (as measured in kilobytes of memory) and for the distribution of those capabilities across functions. In the federated example, a dedicated stand–alone processor containing the processing capacity required is included for each function. In the integrated case two large processors support all avionics functions.

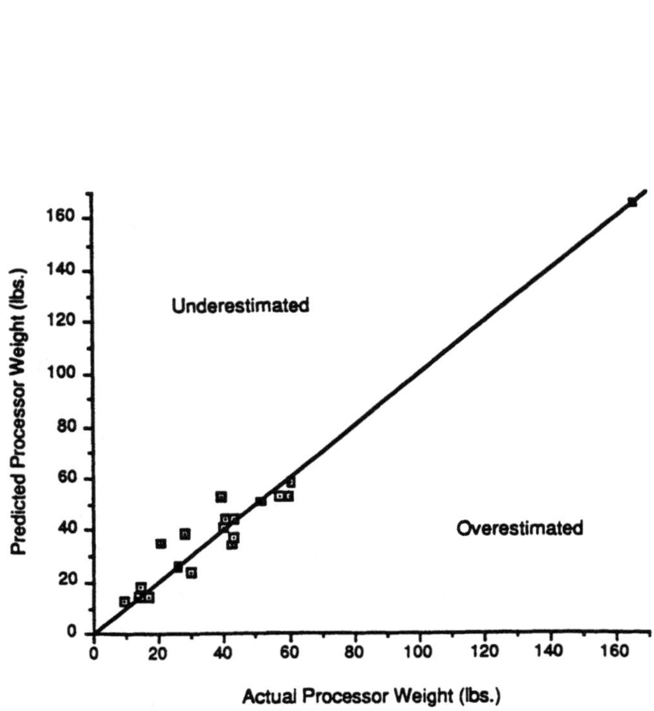

Figure 1–1 Predicted Versus Actual Processor Weight

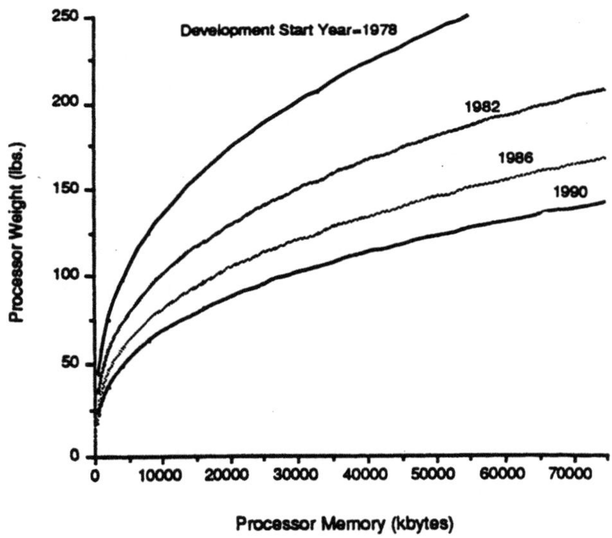

Figure 1–2 Processor Weight as a Function of Capacity and Time

After we assigned estimates of capacity for the individual processors for the two architectures described above, we estimated the weights of the processors using our weight equation, assuming an FSD start date of 1990 for the ATF-generation system. We included only those functions that would most likely be consolidated in integrated processors. Functions that would have their own processors regardless of architecture include flight control and displays. The results of this analysis are included in Table 1-2. The weight estimates from the model are very close to ATF program office engineering estimates for ATF integrated processors.

Table 1-2 Estimated Processor Weight

	CAPACITY (KILOBYTES)	ESTIMATED WEIGHT (POUNDS)
Federated System Processor Function		
Mission Processing	33,000	107
Fire-Control Radar	19,800	89
IRST	16,500	83
Electronic Warfare	13,200	76
Navigation	8,250	64
Communication and Identification	16,500	83
Digital Map	57,750	131
Total	165,000	633
Integrated System		
Processor 1	65,000	137
Processor 2	100,000	160
Total	165,000	297
Weight Difference	-	336

The next step was to generate estimates of aircraft weight and engine thrust for both the integrated and federated examples. In this exercise the integrated system served as the baseline. We started with a baseline gross takeoff weight (GTOW) of 60,000 pounds. Using an engineering rule-of-thumb derived from equations in [1] (indicating that an increase in 1 pound of avionics weight results in an increase in 7.2 pounds of GTOW), we estimated a federated system GTOW. As most cost-estimating relationships (CERs) use empty weight or airframe unit weight (AUW) as inputs, we used existing statistical models to derive these values from GTOW. Holding mission capability constant between the two configurations, an increase in GTOW would also dictate an increase in maximum engine thrust. To estimate engine thrust we assumed a twin-engine configuration and a thrust–to–weight ratio at GTOW of 1.2. Table 1–3 lists differences in aircraft weight and thrust between hypothetical aircraft with integrated and federated avionics systems.

Table 1–3 Hypothetical Aircraft Weight Differences

	Integrated	Federated	Difference
Gross Takeoff Weight (pounds)	60,000	62,404	2,404
Weight Empty (pounds)	38,790	40,327	1,536
Aircraft Unit Weight (pounds)	26,313	27,207	894
Engine Thrust (per engine, pounds)	36,000	37,443	1,443

Once we derived aircraft thrust and weight estimates for both integrated and federated systems, we used these values as inputs into CERs taken from [2], [3], [4], and [5] predicting airframe and engine development, production, and O&S costs. Note that avionics costs are assumed to be the same for both configurations. The program assumptions used for our cost estimates included a production run of 1,000 aircraft with 750 aircraft in the operational inventory for 20 years.

Table 1–4 presents our cost analysis results comparing costs of integrated and federated systems. All costs are in millions of FY 1990 dollars. Our analyses indicate a total program savings of 1.9 billion dollars. None of these cost estimates should be considered an estimate for any specific future system. The estimates are simply a device to arrive at a credible approximation of system cost savings attributable to integrated avionics for some typical future system.

In addition to life–cycle cost savings, non–pecuniary benefits are also attributable to integrated avionics. Combining the processing in an integrated architecture increases the potential functionality of the avionics suite. That is, having the software for the various subsystems execute in a common processing area provides the opportunity for the software to do more. System reconfiguration and sensor fusion are two additional functions that result from the integrated architecture. Implementing these functions requires additional software as well as increasing the complexity of integrating the software. The additional software and complexity increase the life–cycle cost, while at the same time the increased functionality increases system effectiveness.

Although these additional functions will increase software life–cycle costs, they will also improve the overall system effectiveness of the avionics suite. The improvement in system

Table 1–4 Cost Savings From Integration

	Millions of FY 1990 Dollars		
	Integrated	Federated	Difference
Airframe and Engine Development			
Airframe Development	$4,852	$4,971	$118
Engine Development	3,462	3,528	66
Total	$8,314	$8,499	$185
Airframe Production			
Cumulative Airframe Cost	$23,028	$23,694	$666
Airframe Cumulative Average Cost	$23.0	$23.7	$.7
Engine Production (two Engines per Aircraft)			
Cumulative Engine Cost	$9,760	$10,064	$304
Cumulative Average Engine Cost	4.9	5.0	.2
Engine and Airframe Development and Production	$41,103	$42,257	$1,154
Operating and Support Costs Affected by Integration			
Fuel Cost per Year per Aircraft	$.667	$.694	$.027
Fuel Cost, 20 Year Total	10,000	10,408	408.1
Airframe O&S per Year per Aircraft	.535	.539	.004
Airframe O&S, 20-Year Total	8,019	8,086	66.1
Engine O&S per Year per Aircraft	1.025	1.042	.016
Engine O&S, 20-Year Total	15,378	15,624	246.4
O&S Affected by Integration per Year per Aircraft	2.226	2.275	.048
O&S Affected by Integration, 20 Year Total	33,397	34,118	721
Total Program Savings from Integration			$1,875

effectiveness or performance is difficult to quantify. Improved performance can be characterized in three ways: (1) increased probability of returning to base; (2) increased probability of kill; and (3) increased sortie rate. The three measures are interrelated. To quantify their potential benefits and assign some pecuniary value to them is beyond the scope of this paper.

2. SOFTWARE COSTS OF INTEGRATION

In this section we attempt to quantify the increase in life–cycle costs of the software in an integrated architecture. This is done for physical and functional integration, where physical integration is simply the consolidation of processing capability into one or two large integrated processors. When we refer to physical integration we imply no additional functionality over the baseline federated system. Functional integration involves letting the software perform additional functions not provided for in the federated system, where physical integration is a necessary condition for functional integration.

The impact of an integrated avionics suite on software life–cycle cost results from changes in software size and complexity. Our approach attempts to measure the change in software size for inputting in CERs, and to adjust the CERs to reflect changes in complexity. We examine three phases of the life cycle: development, integration, and O&S.

We used CERs presented in [6] that estimate development effort, measured in man months (MM), as a function of size, measured in thousands of lines of code (KLOC). The relationships are of an exponential form. For the operational flight program (OFP) the relationship is:

$$MM = 5.8\ (KLOC)^{1.21}$$

For the support software the relationship is:

$$MM = 4.18\ (KLOC)^{1.05}$$

The cost effects of changes in size resulting from integrated avionics is captured directly in the development cost relationships. Estimated development cost in turn drives integration and O&S costs. Integration costs are estimated as a percentage of the development cost, where the percentage ranges from 27% to 40%, depending on the complexity of the software. The O&S costs in man months are estimated as a percentage of the development plus integration costs. In our analyses we use a "change traffic" factor of 0.15 operations per year for this estimate.

Changes in complexity will be reflected in the integration and O&S cost–estimating relationships. Complexity is not reflected in the development CER since each subsystem's software is physically located on its own processing module within a common processing area. Therefore, in terms of complexity, developing software for a physically integrated avionics suite is basically the same as for a federated avionics suite. This is true for functional integration also. The complexity of functionally integrating the software shows up when integrating the subsystems together and then maintaining the more complex software.

As noted above, complexity is handled in the integration cost by varying the percentage of integration cost in development cost. The more complex, the greater the percentage. For O&S cost, complexity is reflected in the mix of organic and contractor personnel maintaining the software in the "logistics segment" of O&S (defined as the final 15 years of operations). The more complex the software, the greater the percentage of contractor personnel. Contractor personnel are used exclusively in the "production segment" defined as the first 5 years of operations.

The estimated number of MM are converted to dollars using one of two dollar–per–MM ($/MM) rates. For development, integration, and O&S contractor personnel the rate of $12,500/MM is used. For O&S organic personnel the rate of $4,160/MM is used. These rates were provided by Air Force Logistics Center personnel at Warner Robbins Air Force Base.

The change in software size resulting from transitioning to an integrated avionics suite is dependent on the degree of implementation of additional functionality. The increase in software size is in both the OFP and the support software associated with it. The OFP software implements the functionality, while the support software provides the test and simulation software for evaluating the OFP software. The more complicated the OFP software is, the more test software is required to verify the correctness of the OFP software.

For the purpose of the analysis we have defined five different avionics suites. The first is a federated suite that we use as a baseline for all comparisons. The second suite is a physically integrated avionics suite, holding functionality constant at the federated avionics suite level. The last three suites are all physically integrated suites with varying degrees of functionality introduced. These three suites are designated low, medium, and high, depending on the degree of functional integration.

The differences in functionality between the low, medium, and high cases revolves around the degree of implementation of the function's system reconfiguration and sensor fusion. Varying degrees of system reconfiguration can be simple redundancy or the more complicated resource reallocation. For sensor fusion the difference in implementation can be between simple sensor cueing or varying degrees of track file improvement, false alarm rate reduction, and pilot workload reduction. For example, the low case may involve both simple redundancy and sensor cueing, while the high case may involve both resource reallocation and sensor fusion algorithms, including pilot workload reduction similar to a Pilots Associate program. The medium case may involve some combination of the two levels of implementation.

Table 2–1 presents information characterizing federated and integrated avionics suite software efforts. Included are size estimates disaggregated by subsystem. As flight control and display processing is not affected by integration, it is ignored in our analyses. The total size

Table 2–1 Software Parameters

		INTEGRATED			
			FUNCTIONAL		
	FEDERATED	PHYSICAL	LOW	MEDIUM	HIGH
Lines of Code (KLOC)					
OFP	950	1,000	1,095	1,331	1,745
Op System, Chip Level	0	50	100	115	115
Mission Processing	200	300	325	400	800
Radar	250	225	230	282	285
Infra-Red	200	175	180	222	225
Electronic Warfare	150	125	130	156	160
Communication, Identification, and Navigation	150	125	130	156	160
Support Software	4,000	4,500	5,000	7,000	10,000
O&S Cost Factors (contractor %, logistics segment)	0.25	0.25	0.35	0.50	0.60
Integration/Rates (% of development cost)					
OPF Software	0.27	0.27	0.30	0.35	0.40
Support Software	0.27	0.27	0.27	0.27	0.27

estimates for both OFP and support software are within the ranges provided by the ATF program office. The break down by subsystem was done by the authors based on discussions with contractors and analogy to previous avionics suites.

The level of functionality is held constant between the federated suite and the physically integrated suite. For functionally equivalent systems the differences in size estimates are small, less then 5% for the OFP software and about 12% for the support software. There is no difference in complexity as reflected in differences in the integration/development cost ratio and the contractor/organic personnel mix for O&S.

Changes in both size and complexity occur with increases in functional implementation. Using the three designated cases of functional integration (low, medium, and high), we vary the size of both the OFP and support software. Again examining Table 2–1, we can see the changes in size estimates for the three cases. Also in Table 2–1, we can see how the increased complexity resulting from the increased functionality translates into increases in the integration/development cost ratio and the contractor/organic personnel mix.

The cost estimates using the values in Table 2–1 are presented in Table 2–2. Included in Table 6 are the development, integration, and O&S cost for both the OFP and support software,

Table 2-2 Software Costs of Integration

		MILLIONS OF FY 1990 DOLLARS							
		INTEGRATED				DIFFERENCES FROM FEDERATED			
			FUNCTIONAL				FUNCTIONAL		
	FEDERATED	PHYSICAL	LOW	MEDIUM	HIGH	PHYSICAL	LOW	MEDIUM	HIGH
OFP									
Development	$208	$219	$242	$307	$445	$10	$34	$99	$236
Integration	56	59	73	107	178	3	16	51	122
O&S	496	520	637	932	1,494	25	141	436	998
Total	$760	$798	$951	$1,346	$2,116	$38	$191	$586	$1,356
Support									
Development	$316	$358	$400	$569	$828	$42	$84	$253	$512
Integration	85	97	108	154	224	11	23	68	138
O&S	753	852	1,028	1,627	2,524	99	275	874	1,770
Total	$1,155	$1,307	$1,536	$2,350	$3,575	$152	$381	$1,195	$2,420
Totals									
Development	$525	$577	$642	$876	$1,273	$52	$117	$352	$748
Integration	142	156	181	261	401	14	39	119	260
O&S	1,249	1,373	1,665	2,559	4,017	124	416	1,310	2,768
Total	$1,915	$2,105	$2,487	$3,696	$5,691	$190	$572	$1,781	$3,776

for each of the five avionics suites. Also included are the differences between the integrated cases and the federated case.

As can be seen from the table the difference in cost between the federated and integrated avionics suite, holding functionality constant, are small, about 10%. The difference increases as the functionality increases, to about 200%. The estimates presented here do not represent estimates for any specific program. The estimates represent only our interpretation of the various options available and the possible differences in cost that could exist.

3. CONCLUSIONS

A comparison of the differences in software life-cycle cost with the cost savings in hardware life-cycle cost reveals that, when there is no increase in functionality, there are clear savings with an integrated avionics suite. As functionality of the suite increases, software life-cycle cost increases and can be greater than the costs savings, dependent on the functionality implemented. Table 3-1 presents net integration benefits for different levels for avionics suite funtionality, where the suite with only physical integration is functionally equivalent to the federated suite. With increases in functionality come greater system effectiveness, a more

Table 3-1 **Net Integration Benefits**

	MILLIONS OF FY 1990 DOLLARS			
	PHYSICAL INTEGRATION	LEVEL OF FUNCTIONAL INTEGRATION		
		LOW	MEDIUM	HIGH
Development (with integration)	$119	$29	$-286	$-823
Production	970	970	970	970
O&S	597	305	-589	-2,047
Total	$1,686	$1,304	$95	$-1,900

difficult term to quantify for evaluation. This increase in functionality would not be practicably attainable with a federated avionics suite.

REFERENCES

1. Roskam, Jan, *Airplane Design Part I: Preliminary Sizing of Airplanes*, Roskam Aviation and Engineering Corporation, 1989.

2. Harmon, Bruce R., J. Richard Nelson, Mitchell S. Robinson, Kathryn L. Wilson, Steven R. Shyman, *Military Tactical Aircraft Development Costs*, 5 volumes, Institute for Defense Analyses, Report R–339, September 1988.

3. Goree, Paul F., Joseph S. Domin, Philip A. Evans, Michael C. Frieders, Timothy J. Graves, Thomas A. Musson, Terri J. Walsh, Bruce R. Harmon, *A User's Guide for the Force Acquisition Cost System (FACS)*, Institute for Defense Analyses, Paper P–2550, February 1991.

4. Harmon, Bruce R., Scot A. Arnold, and J. Richard Nelson, *Unit Cost Implications of New Materials: Preliminary Analyses of Airframe Experience*, Institute for Defense Analyses, Document D–908, forthcoming.

5. Northrop Corporation Aircraft Division, *Aircraft Operating and Support Cost Models for Advanced Design*, January 1988.

6. Shyman, Steven R. and T. Keith Blankenship, *Improving Methods for Estimating Software Development Cost*, Institute for Defense Analyses, Paper P–2388, April 1990.

TRANSFORM TECHNIQUES IN COST ANALYSIS

by

John A. Long
TASC

ABSTRACT

Risk is a fact of life that must be addressed by cost analysts. The analyst cannot eliminate the risk but can, at least, capture, present, and explain the aspects of risk to the decision maker. Risk analysis is a procedure for analyzing how randomness affects the cost. To place a cost estimate in proper perspective, it must be viewed as a random variable produced by other random variables.

Cost analysts have used a variety of techniques including Monte Carlo simulation, additive and multiplicative moments, and simple sums and products of random variables to estimate risk. This paper applies the method of transformation to this problem.

Basic Cost Estimating Models — The models or mathematical expressions used in cost estimating appear in two general forms: the additive model and the multiplicative model. The additive model is expressed as

$$Y = X_1 + X_2 \tag{1}$$

and the multiplicative model as

$$Y = X_1 X_2 \tag{2}$$

where X_1 and X_2 are random variables. The mathematical expressions used in practice may appear to be more complex, but are usually reducible to these two general forms. These models are referenced frequently in the following discussion.

Method of Transforms — In the same way that logarithms are used to manipulate complicated expressions in elementary algebra, transforms are used to simplify operations in the algebra of random variables. The use of transforms requires no more assumptions and little more knowledge than other methods, yet affords an added bonus in that the resultant expression can be used to determine the exact distribution of the sum or product of random variables. If the analyst is only concerned with the moments of that distribution, they are obtainable directly from the transform without going through the sometimes difficult task of transform inversion.

Three transform types will be discussed in this paper. These are the Laplace transform for the additive model (1), the Mellin transform for the multiplicative model (2), and the geometric, in conjunction with the Laplace, for a special case of the additive model.

The transform most used in finding the distribution of the sum of continuously distributed non–negative random variables is the Laplace transform. Given a continuous random variable x, ($x \geq 0$), with probability density function f (x).

$$L[f(x)] = \int_0^{\infty} e^{-sx} f(x)\, dx \tag{3}$$

is the unilateral Laplace transform of f(x), where s is a complex variable of the form $s = a + iw$. Its inversion integral is

$$L^{-1}[f(x)] = f(x) = (2\pi i)^{-1} \lim_{b \to \infty} \int_{a-ib}^{a+ib} e^{sx} L[f(x)]\, ds \tag{4}$$

Equations (3) and (4) above constitute the Laplace transform pair. Note that the inversion requires the evaluation of a contour integral. This complex evaluation can be avoided by the use of tables of transform pairs, such as those found in Giffin [3] and Springer [5].

In multiplying random variables, the Mellin transform is used to develop product probability distributions. The Mellin transform for the probability density function f(x) of the random variable x is

$$M[f(x)] = \int_0^{\infty} x^{s-1} f(x)\, dx \tag{5}$$

where s is once again a complex variable. The Mellin inversion integral is

$$M^{-1}[f(x)] = f(x) = (2\pi i)^{-1} \lim_{b \to \infty} \int_{a-ib}^{a+ib} x^{-s} M[f(x)] \, ds \qquad (6)$$

Equations (5) and (6) above are the Mellin transform pair. Again, the inversion requires the evaluation of a contour integral. As in the case of the Laplace transform, this integration can be avoided by using tables such as those in Giffin [3] and Springer [5].

In order to properly exercise the additive and multipicative models, the analyst must apply the convolution operation. Consider the additive model (1), where X_1 and X_2 are continuous independent random variables with probability density functions $f_1(x_1)$ and $f_2(x_2)$, respectively. The distribution of Y is given by the convolution integral

$$g(Y) = \int_0^Y f_1(X) \, f_2(Y-X) \, dx = f_1 * f_2 \qquad (7)$$

where the symbol "*" represents the convolution operation. In the transform domain, the convolution of two functions, according to Borel's theorem, is equal to the inverse of the product of their transforms. Simply stated in terms of the Laplace transform.

$$L[g(Y)] = L[f_1(x_1)] \, L[f_2(x_2)] \qquad (8)$$

or the transform of the probability density function of a sum is the product of the transforms of the density functions of the individual random variables being added. This procedure can be logically extended to the sum of n continuous independent random variables

$$Y = X_1 + X_2 + \ldots + X_n \qquad (9)$$

as

$$L[g(Y)] = \prod_{i=1}^{n} L[f_i(x_i)] \qquad (10)$$

where π is the multiplication operator. If the n terms in the sum are all identically distributed independent random variables, (10) reduces to

$$L[g(Y)] = (L[f(x)])^n \tag{11}$$

If the inversion of the transform of g(Y) proves to be too difficult and the analyst is only interested in the moments of the distribution of g(Y), these can be obtained rather easily. In the case of the Laplace transform, the moments of the distribution can be determined by simply taking derivatives of the transform with respect to s and evaluating at $s=0$ [3:62–63].In mathematical terms,

$$d^rL[f(x)]/ds^r\Big|_{s=0} = \int_0^\infty (-x)^r e^{-sx} f(x)\, dx\Big|_{s=0} \tag{12}$$

and the moments about the origin can be expressed as

$$E[x^r] = (-1)^r\, d^rL[f(x)]/ds^r\big|_{s=0} \tag{13}$$

If one is interested in the moments about the mean, μ, the logic above can be extended. That is,

$$E[(x-\mu)^n] = (-1)^n\, d^n/ds^n(e^{su}\; L[f(x)])\big|_{s=0} \tag{14}$$

Thus, the variance and higher order moments about the mean can be easily and quickly determined given the distribution function of the random variable x.

Now, consider a special case of the Laplace transform where the number of terms in the sum is itself a random variable. This situation is conveniently handled through the application of the geometric transform.

The geometric transform is the discrete analog of the Laplace transform and provides the same capabilities with respect to discrete random variables as the Laplace with continuous random variables. The geometric transform is defined by the sum

$$G[p(n)] = \sum_{n=0}^{\infty} p(n)z^n = G_n(z) \tag{15}$$

where p(n) is the probability mass function of n and z is a complex transform variable corresponding to s in the Laplace transform. Thus, the geometric transform of the function p(n) can be viewed as the expected value of z^n. The expression in (9) can then be treated as a conditional transform. That is, $L[g(Y/n)]$ is the transform of $g(Y)$ for some fixed n. The unconditional transform is then obtained by weighting each term by $p(n)$ and summing over all n. That is,

$$L[g(Y)] = \sum_{n=0}^{\infty} L[g(Y)/n)]p(n) = \sum_{n=0}^{\infty} (L[f(x)]^n p(n) \tag{16}$$

Note that this expression closely resembles that of the geometric transform (15) with the z replaced by $L[f(x)]$. The transform for the random sum of random variables is then expressed by

$$L[g(Y)] = G_n(z)|_{z=L[f(x)]} = G_n[L[f(x)]] \tag{17}$$

Although this expression appears to be a geometric transform, it is not. The end result is a Laplace transform in the complex variable s.

Often in cost analysis problems, historical data or subjective estimates are so sparse or incomplete as to permit only an estimate of the mean and variance of the random variables x and n. In those cases where there is insufficient, credible data to fit distributions to the random variables involved, one can still solve for the mean and variance of the random sums expressed in equation (9). That is,

$$\mu_Y = \mu_n \mu_x \tag{18}$$

and

$$\sigma_Y^2 = \mu_n \sigma_x^2 + \mu_x^2 \sigma_n^2 \tag{19}$$

Thus, given only the mean and variance of both the number of terms and the individual terms in the sum, the analyst can compute the mean and variance of the random sum. The analyst need not know the distribution of the transform to obtain these important statistics. Equations (18) and (19) hold for both continuous and discrete X.

The Mellin convolution used in the multiplicative model takes a slightly different form. The probability density function of the product of two non–negative, continuously distributed, independent random variables model (2), with probability density functions $f_1(x_1)$ and $f_2(x_2)$, is in terms of the Mellin convolution,

$$g(Y) = \int_0^\infty 1/x \, f_1(Y/x) \, f_2(x) \, dx \quad (20)$$

Then, the transform of the Mellin convolution is

$$M[g(Y)] = M[f_1(x_1)] \, M[f_2(x_2)] \quad (21)$$

where M represents the Mellin transform of the densities involved. In parallel with the sum of n continuous independent random variables given above, the transform of the density function of the product of n non–negative independent random variables is

$$M[g(Y)] = \prod_{i=1}^{n} M[f_i(x_i)] \quad (22)$$

The recovery of the moments of the Mellin transform is even easier than the Laplace and does not involve derivatives but a mere substitution of variables. It can be shown that the rth moment about the origin can be derived by replacing the complex variable s in the Mellin transform with $s = r+1$ [3:74]. In mathematical symbols,

$$E[x^r] = M[f(x)]\big|_{s=r+1} = \mu_r' \quad (23)$$

If the analyst is interested in central moments rather than moments about the origin as expressed in (23), the conversions given in Table 1 can be used. Table 2 is a compilation of common probability distributions and their transforms.

Table 1 Conversion – Origin to Central Moments

$\mu_1 = \mu_1'$
$\mu_2 = \mu_2' - (\mu_1')^2$
$\mu_3 = \mu_3' - 3\mu_1'\mu_2' + 2(\mu_1')^3$
$\mu_4 = \mu_4' - 4\mu_1'\mu_3' + 6(\mu_1')^2\mu_2' - 3(\mu_1')^4$

This discussion of transforms has thus far been limited to the case where the random variables on the right hand side of the equation are statistically independent. Information relative to the statistically dependent case is rather lacking. This is primarily due to the fact that the

Table 2 Distributions and Transforms

NAME	DENSITY FUNCTION-f(x)	MEAN	VARIANCE	LaPLACE TRANSFORM	MELLIN TRANSFORM
Rectangular	$\frac{1}{b-a}$, $a \leq x \leq b$	$\frac{a+b}{2}$	$\frac{(b-a)^2}{12}$	$\frac{e^{-sa}-e^{-sb}}{s(b-a)}$	$\frac{b^s-a^s}{s(b-a)}$
Exponential	ae^{-ax} , $x > 0$	$\frac{1}{a}$	$\frac{1}{a^2}$	$\frac{a}{s+a}$	$(\frac{1}{a})^{s-1}$ $\Gamma(s)$
Gamma	$\frac{b(bx)^{a-1} e^{-bx}}{\Gamma(a)}$ $a \leq x < 00$, $a > 0$, $b > 0$	$\frac{a}{b}$	$\frac{a}{b^2}$	$(1 + \frac{s}{b})^{-a}$	$b^{-s+1} \frac{\Gamma(s+a-1)}{\Gamma(a)}$
Beta	$\frac{\Gamma(a+b)}{\Gamma(a)\,\Gamma(b)} \chi^{a-1} (1-\chi)^{b-1}$ $0 < \chi < 1$, $a > 0$, $b > 0$	$\frac{a}{a+b}$	$\frac{ab}{(a+b)^2(a+b+1)}$	$\frac{\Gamma(a+b)}{\Gamma(a)} \sum_{j=0}^{\infty} (-1)^j \frac{\Gamma(a+j)s^j}{\Gamma(a+b+j)\,(j+1)}$	$\frac{\Gamma(a+b)}{\Gamma(a)} \frac{\Gamma(s+a-1)}{\Gamma(s+a-1+b)}$
Standardized Normal	$\frac{1}{\sqrt{2\pi}} \exp \left[\frac{-\chi^2}{2}\right]$ $-\infty < \chi < \infty$	0	1	$\exp(s^2/2)$	$\frac{(2)^{(s-3)/2}}{\sqrt{\pi}}$ $\Gamma(s/2)$
	MASS FUNCTION-f(x)			GEOMETRIC TRANSFORM	
Binomial	$\binom{n}{\chi} p^{\chi} (1-p)^{n-\chi}$	np	np(1-p)	$[pz + (1-p)]^n$	
Poisson	$u^{\chi} e^{-u}/\chi!$	u	u	$e^{-u(1-z)}$	
Geometric	$P(1-p)^{\chi}$	$\frac{1-p}{p}$	$\frac{1-p}{p^2}$	$P/[1-(1-p)z]$	

analysis is considerably more complicated. For the additive model (1), the transform of the probability density function of the sum of dependent random variables is no longer the product of the transforms of the probability density functions of the component random variables [5:67]. Springer [5] uses the Fourier transform or multivariate characteristic function to evaluate the sum of dependent random variables. To apply the technique, the analyst must know the multivariate probability density functions.

For the multiplicative model (2), Springer [5] uses the two dimensional Mellin transform to evaluate the product of two dependent random variables. Again, the analyst must know the multivariate probability density function.

Thus, the fundamental problem in the dependent case is that evaluation of both the additive and multiplicative models requires a knowledge of the multivariate probability density function. Given the present state of the art in cost estimating, it seems unlikely that these distributions can be identified with any degree of confidence. The consequence of ignoring dependence is an error in the variance for the additive model and an error in both the expected value and variance for the multiplicative model. The magnitude of the error is a function of the magnitude of the covariance and, where applicable, the expected value of the dependent variables involved.

Application of Transforms — Consider a simple problem in cost analysis where the analyst needs to compute the total cost of two components each of which is independent and identically distributed following the negative exponential distribution. That is,

$$y = x_1 + x_2 \tag{24}$$

where

$$f(x_i) = ae^{-ax_i}, \qquad 0 \leq x_i < \infty \tag{25}$$

The Laplace transform of the individual cost is

$$L\,[f(x_i)] = \int_0^\infty ae^{-ax_i}\, e^{-sx_i}\, dx_i = \frac{a}{s+a}$$

using equation (3) and Table 2.

Since x_1 and x_2 are identically distributed, the convolution theorem (7) yields

$$L[f(y)] = \{L[f(x_i)]\}^2 = \left[\frac{a}{s+a}\right]^2 \tag{26}$$

Equation (26) could be inverted to find the distribution function of x or a table look up method could be used to find the distribution. However, the analyst is primarily interested in the expected value and variance of that distribution in order to do a risk analysis. Therefore, applying equation (13),

$$E[y] = (-1)\frac{dL[f(y)]}{ds}\bigg|_{s=0} = \frac{2a^2}{(s+a)^3}\bigg|_{s=0} = \frac{2}{a} \tag{27}$$

The variance of x using the second formula in Table 1 is expressed as

$$\sigma_y^2 = E[y^2] - \{E[y]\}^2 \tag{28}$$

The second moment of the total cost distribution is

$$E[y^2] = \frac{d^2L[f(y)]}{ds^2}\bigg|_{s=0} = \frac{6a^2}{(s+a)^4}\bigg|_{s=0} = \frac{6}{a^2}$$

and the variance applying equation (28) is

$$\sigma_y^2 = E[y^2] - \{E[y]\}^2 = \frac{6}{a^2} - \frac{4}{a^2} = \frac{2}{a^2} \tag{29}$$

Taking the above example one step further, consider the case where the analyst must compute the cost of n components each of which is independent and identically distributed following the negative exponential distribution where n is a random variable from the Poisson distribution. That is,

$$y = x_1 + x_2 \ldots + x_n \tag{30}$$

As before,

$$L\,[f(x_i)] = \int_0^{\infty} ae^{-ax_i}\, e^{-sx_i}\, dx_i = \frac{a}{s+a}$$

As for n, the Poisson is a discrete distribution and, therefore, the geometric transform must be applied as in equation (15). Since

$$p(n) = \frac{\lambda^n}{n!}\, e^{-\lambda} \qquad \text{where } n = 0, 1, 2, \ldots \tag{31}$$

the transform is

$$G[p(n)] = \sum_{n=0}^{\infty} p(n)z^n = e^{-\lambda(1-z)} \tag{32}$$

From equation (17),

$$L[g(y)] = G_n\{L[f(x)]\} = \exp\left[-\lambda\left(1-\frac{a}{s+a}\right)\right] = \exp\left(-\frac{\lambda s}{s+a}\right) \tag{33}$$

Note that the end result is a Laplace transform. The moments of the distribution of g(y) could be extracted using (13), as in the previous example, or equations (18) and (19) could be applied yielding

$$\mu_y = \mu_n\mu_x = \lambda(1/a) = \frac{\lambda}{a} \tag{34}$$

and

$$\begin{aligned} \sigma_y^2 &= \mu_n\sigma_x^2 + \mu_x^2\sigma_n^2 \\ &= \lambda\frac{1}{a^2} + \left(\frac{1}{a}\right)^2 \lambda \\ &= \frac{2\lambda}{a^2} \end{aligned} \tag{35}$$

As a final example and one using the Mellin transform, consider the situation where the analyst has two random variables, x and y, distributed rectangularly and wants to know the distribution, mean, and variance of the product

$$z = xy \tag{36}$$

The Mellin transform from Table 2 for the rectangular distribution is

$$M[f(x)] = \int_0^\infty x^{s-1} f(x)\, dx =$$

$$= \frac{b^s - a^s}{s(b-a)} \tag{37}$$

where

$$f(x) = \frac{1}{b-a} \qquad \text{and } a \leq x \leq b \tag{38}$$

If f(x) from equation (36) has parameters $a \leq x \leq b$ and f(y) has parameters $c \leq y \leq d$, then the Mellin convolution for f(z) becomes

$$M[f(z)] = M[f(x)]\, M[f(y)]$$

$$= \frac{b^s - a^s}{s(b-a)} \frac{d^s - c^s}{s(d-c)}$$

$$= \frac{(b^s-a^s)\,(d^s-c^s)}{s^2\,(b-a)\,(d-c)} \tag{39}$$

This transform can then be inverted using (6) to yield the exact density function for the product of two rectangularly distributed random variables. However, once again, the analyst is primarily interested in the moments of this distribution. Using equation (23),

$$E[(z)] = M[f(z)]|_{s=2}$$

$$= \frac{(b^2-a^2)\,(d^2-c^2)}{4(b-a)\,(d-c)} \tag{40}$$

The second moment about the origin is

$$E[(z^2)] = M[f(z)]|_{s=3}$$

$$= \frac{(b^3-a^3)\,(d^3-c^3)}{9(b-a)\,(d-c)} \tag{41}$$

and thus the variance of f(z) is expressed as

$$\sigma_z^2 = E[z^2] - \{E[z]\}^2$$

$$= \frac{(b^3-a^3)\,(d^3-c^3)}{9(b-a)\,(d-c)} - \left[\frac{(b^2-a^2)\,(d^2-c^2)}{4(b-a)\,(d-c)}\right]^2 \tag{42}$$

Although this is a formidable looking expression, with real values substituted for a, b, c, and d, the evaluation is a matter of simple arithmetic.

Conclusion — The use of transform techniques in cost analysis is but one way for the analyst to manipulate random variables and perform risk analysis. It is important to note that in using transforms, the calculations are exact and not approximate. Therefore, the analyst introduces no errors into the analysis when using these techniques. For further, indepth reading on this subject, the analyst is directed to the books by Giffin and Muth as listed in the bibliography.

BIBLIOGRAPHY

1. Doetsch, Gustav, *Guide to the Applications of the Laplace and Z–Transforms,* Van Nostrand Reinhold Company, London, 1971.
2. Giffin, Walter C., *Introduction of Operations Engineering,* Richard D. Irwin, Inc., Illinois, 1971.
3. Giffin, Walter C., *Transform Techniques for Probability Modeling,* Academic Press, New York, 1975.
4. Muth, Eginhard J., *Transform Methods with Applications to Engineering and Operations Research,* Prentice–Hall, Inc., Englewood Cliffs, New Jersey, 1977.
5. Springer, M. D., *The Algebra of Random Variables,* John Wiley & Sons, New York, 1979.
6. Willis, Raymond E. and Norman L. Chervany, *Statistical Analysis and Modeling for Management Decision–Making,* Wadsworth Publishing Company, Inc., Belmont, California, 1974.

LEARNING CURVE AND RATE ADJUSTMENT MODELS: COMPARATIVE PREDICTION ACCURACY UNDER VARYING CONDITIONS

O. Douglas Moses
Associate Professor
Department of Administrative Science
Naval Postgraduate School
Monterey, CA 93943

ABSTRACT

Learning curve models have gained widespread acceptance as a technique for analyzing and forecasting the cost of items produced from a repetitive process. Considerable research has investigated augmenting the traditional learning curve model with the addition of a production rate variable, creating a rate adjustment model. This study compares the predictive accuracy of the learning curve and rate adjustment models. A simulation methodology is used to vary conditions along seven dimensions. Forecast errors are analyzed and compared under the various simulated conditions, using ANOVA. As in all simulation studies, findings must be interpreted in light of the assumptions underlying the simulation. Overall results indicate that neither model dominates; each is more accurate under some conditions. Conditions under which each model tends to result in lower forecast errors are identified and discussed. This work was sponsored by the Cost Estimating and Analysis Division of the Naval Sea Systems Command and the Naval Postgraduate School.

1. INTRODUCTION

Learning curves have gained widespread acceptance as a tool for planning, analyzing, explaining, and predicting the behavior of the unit cost of items produced from a repetitive production process. (See Yelle, 1979, for a review.) Cost estimation techniques for planning the cost of acquiring weapon systems by the Department of Defense, for example, typically consider the role of learning in the estimation process. The premise of learning curve analysis is that cumulative quantity is the primary driver of unit cost. Unit cost is expected to decline as cumulative quantity increases.

There is general acknowledgement that cumulative quantity is not the only factor that influences unit cost and that the simple learning curve is not a fully adequate description of cost behavior. Hence prior research has attempted to augment learning curve models by including additional variables (e.g., Moses, 1990). Most attention has been focused on the addition of a production rate term.[1] The resulting augmented model is usually referred to as a rate adjustment model.

Conceptually, production rate should be expected to affect unit cost because of the impact of economies of scale. Higher production rates may lead to several related effects: greater specialization of labor, quantity discounts and efficiencies associated with raw materials purchases, and greater use of facilities permitting fixed overhead costs to be spread over a larger output quantity. Together, these effects work to increase efficiency and reduce production cost (Bemis, 1981; Boger and Liao, 1990; Large, et. al., 1974; Linder and Wilbourn, 1973). However, higher production rate does not guarantee lower cost. When production rate exceeds capacity, such factors as over–time pay, lack of skilled labor, or the need to bring more facilities online may lead to inefficiencies and increased unit cost. In short, production rate may be associated with both economies and diseconomies of scale.

2. PRIOR RESEARCH

Numerous studies, using data on actual production cost elements, have been conducted to empirically examine the impact of production rate on unit cost. The broad objective of the research has been to document rate/cost relationships and determine if consideration of production rate leads to improvements in cost explanation or prediction. Results have been inconsistent and general findings inconclusive. Various studies (e.g., Alchian, 1963; Cochran, 1960; Hirsh, 1952; Large, Campbell and Cates, 1976) found little or no significance for rate variables. Other studies did document significant rate/cost relationships (e.g., Bemis, 1981; Cox and Gansler,

[1]One review of the literature pertaining to learning curves (Cheney, 1977) found that 36% of the articles reviewed attempted to augment the learning curve model in some manner by the inclusion of production related variables.

1981). Some research found significant results only for particular individual cost elements, such as labor (Smith, 1976), tooling (Levenson, et. al., 1971) or overhead (Large, Hoffmayer, and Kontrovich, 1974). But rate/cost relationships for these same cost elements were not consistently evident in other studies. When significant, estimates of the rate/cost slope varied greatly and the direction of the relationship was sometimes negative and sometimes positive (e.g., Moses, 1990). In reviewing the existing research on production rate, Smith (1980) concluded that a rate/cost relationship may exist but that the existence, strength and nature of the relationship varies with the item produced and the cost element examined.

Several explanations for these varying, inconclusive empirical results can be offered:

(a) Varying results are to be expected because rate changes can lead to both economies and diseconomies of scale.

(b) Production rate effects are difficult to isolate empirically because of colinearity with cumulative quantity (Gulledge and Womer, 1986).

(c) Researchers have usually used inappropriate measures of production rate leading to misspecified models (Boger and Liao, 1990).

(d) The impact of a production rate change is dominated by other uncertainties (Large, Hoffmayer, and Kontrovich, 1974), particularly by cumulative quantity (Asher, 1956). Alchian (1963), for example, was unable to find results for rate adjustment models that improved on the traditional learning curve without a rate parameter.

3. OBJECTIVE OF THE STUDY

The prior research suggests that consideration of production rate sometimes improves cost explanation, but not always. The prior research suggests that a traditional learning curve model sometimes is preferable to a rate adjustment model, but not always. The prior research provides little guidance concerning the circumstances under which explicit incorporation of production rate into a learning curve model is likely to lead to improved explanation or prediction. This issue is important in a number of cost analysis and cost estimation situations. Dorsett (1990), for example, describes the current situation facing military cost estimators who, with the military facing budget reductions and program stretchouts, are required to rapidly develop weapon system acquisition cost estimates under many different quantity profiles. One choice the cost analyst faces is between using a rate adjustment model or a traditional learning model to develop estimates.[2]

The objective of this paper is to address the following broad issue: Under what circumstances is it beneficial to explicitly consider production rate in cost analysis? That broad issue is addressed by focusing more narrowly on the following research question: Under what circumstances do rate adjustment models outperform traditional learning curve models? Performance here means the ability to provide accurate predictions of future cost.[3]

4. RESEARCH APPROACH

Operationally the research question implies a comparison of the predictive accuracy of two competing cost estimation models. The two competing models were as follows:

The traditional learning curve model, which predicts unit cost as a function of cumulative quantity:[4]

$$C_L = aQ^b \tag{1}$$

[2]Two other techniques for making cost estimates when production rate changes are also mentioned by Dorsett: curve rotation, which involves an ad hoc upward or downward revision to the slope of the learning curve, and the use of repricing models (e.g., Balut, 1981; Balut, Gulledge, and Womer, 1989) which adjust learning curve estimates to reflect a greater or lesser application of overhead cost. Dorsett criticized curve rotation for being subjective and leading to a compounding of error when the prediction horizon is not short. He criticized repricing models because they must be plant–specific to be effective.

[3]For many of the studies cited in the section on prior research, performance was assessed in terms of a model's ability to ex post, statistically account for the variance in an actual cost series. The problem of "overfitting" a model to data always exists when regression techniques are used to ex post explain a cost series. In fact, if higher R^2 is achieved by overfitting, predictive ability can be reduced (Wetherill, 1986).

[4]Note that this is an incremental unit cost model rather than a cumulative average cost model. Liao (1988) discusses the differences between the two approaches and discusses why the incremental model has become dominant in practice. On reason is that the cumulative model weights early observations more heavily and, in effect, "smooths" away period–to–period changes in average cost.

where

C_L	=	Unit cost of item at quantity Q (i.e., with learning considered)
Q	=	Cumulative quantity produced
a	=	Theoretical first unit cost
b	=	Learning curve exponent (which can be converted to a learning slope by slope = 2^b). The most widely used rate adjustment model, which modifies the traditional learning curve model with the addition of a production rate term:

$$C_R = aQ^bR^c \qquad (2)$$

where

C_r	=	Unit cost of item at quantity Q and production rate per period R (i.e., with production rate as well as learning considered)
Q	=	Cumulative quantity produced
R	=	Production rate per period measure
a	=	Theoretical first unit cost
b	=	Learning curve exponent
c	=	Production rate exponent (which can be converted to a production rate slope by slope = 2^c).

A simulation approach was used to address the research question. In brief, cost series were generated under varying simulated conditions. The learning curve model and the rate adjustment model were separately fit to the cost series to estimate model parameters. The estimated models were then used to separately predict future cost. The relative accuracy of the two alternative future cost predictions was determined. Finally, an analysis (ANOVA) was conducted relating relative prediction accuracy (dependent variable) to the simulated conditions (independent variables).

There are three main benefits gained from the simulation approach. First, factors hypothesized to influence prediction accuracy can be varied over a wider range of conditions than would be encountered in any one (or many) sample(s) of actual cost data. Second, explicit control is achieved over the manipulation of factors. Third, noise caused by factors not explicitly investigated is removed. Hence simulation provides the most efficient way of investigating data containing a wide variety of combinations of the factor levels while controlling for the effects of other factors not explicitly identified.

5. RESEARCH CHOICES

There were five choices that had to be made in conducting the simulation experiment:

(1) The form of the rate adjustment (RA) model whose performance was to be compared to the learning curve (LC) model.

(2) The functional form of the cost model used to generate the simulated cost data.

(3) The conditions to be varied across simulation treatments.

(4) The cost objective (what cost was to be predicted).

(5) The measure of prediction performance.

Items (1), (2), (4), and (5) deal with methodological issues. Item (3) deals with the various conditions simulated. The discussion of item (3) encompasses the major variables examined in the study and hence the major issues addressed. Each item will be discussed in turn.

(1) The Rate Adjustment Model — Various models, both theoretical and empirical, have been suggested for incorporating production rate into the learning curve (Balut 1981; Balut, Gulledge, and Womer, 1989; Linder and Wilbourn, 1973; Smith, 1980, 1981; Washburn, 1972; Womer, 1979). The models vary with respect to tradeoffs made between theoretical completeness and empirical tractability. Equation 2, described above, was the specific rate adjustment model analyzed in this study, for several reasons: First, it is the most widely used rate adjustment model in the published literature. Second, it is commonly used today in the practice of cost analysis (e.g., Dorsett, 1990). Third, in addition to cost and quantity data (needed to estimate any LC model), equation 2 requires only production rate data. Thus equation 2 is particularly appropriate for examining the incremental benefit of attention to production rate. (Other RA models offered in the literature require knowledge of still additional variables.)[5] In short, equation 2 is the most widely applicable and most generally used rate adjustment model.

(2) The Cost Generating Function: A "true" cost function for an actual item depends on the item, the firm, the time period and all the varying circumstances surrounding actual produc-

[5]The equation 2 model is particularly applicable in situations where a cost analyst or estimator does not have ready access to or sufficient knowledge about the cost structure and cost drivers of a manufacturer. Examples include the Department of Defense procuring items from government contractors in the private sector, or prime contractors placing orders with subcontractors.

tion. It is likely that most manufacturers do not "know" the true cost function underlying goods they manufacture. Thus the choice of a cost function to generate simulated cost data is necessarily ad hoc. The objective here was to choose a "generic" cost function which had face validity, which included components (parameters and variables) that were generalizable to all production situations, and which resulted in a unit cost that depended on both learning and production rate factors. The following explanation of the cost function used reflects these concerns.

At the most basic level the cost of any unit is just the sum of the variable cost directly incurred in creating the unit and the share of fixed costs assigned to the unit, where the amount of fixed costs assigned depend on the number of units produced.

$$UC = VC + \frac{FC}{PQ} \tag{3}$$

where

UC	=	Unit cost
VC	=	Variable cost per unit
FC	=	Total fixed costs per period
PQ	=	Production quantity per period.

The original concept of "learning" (Wright, 1936) involved the reduction in variable cost per unit expected with increases in cumulative quantity produced. (By definition, fixed costs are assumed to be unaffected by volume or quantity.) To incorporate the effect of learning, variable cost can be expressed as:

$$VC_Q = VC_1 (Q^d) \tag{4}$$

where

Q	=	Cumulative quantity
VC_Q	=	Variable cost of the Qth unit
VC_1	=	Variable cost of the first unit
d	=	Parameter, the learning index.

Substituting into equation 3:

$$UC_Q = VC_1 (Q^d) + \frac{FC}{PQ} \tag{5}$$

Additionally, assume the existence of a "standard" ("benchmark," "normal," "planned") production quantity per period (PQs). Standard fixed cost per unit (SFC) at the standard production quantity would be:

$$SFC = \frac{FC}{PQ_S} \tag{6}$$

The production rate (PR) for any period can then be expressed as a ratio of the production quantity to the standard quantity:

$$PR = \frac{PQ}{PQ_S} \tag{7}$$

The second term of equation (6) can then be rewritten as:

$$\frac{FC}{PQ} = \frac{SFC}{PR} \tag{8}$$

and equation 5 rewritten as:

$$UC_Q = VC_1\,(Q^d) + SFC\,(PR^{-1}) \tag{9}$$

In this formulation it can be seen that total cost per unit is the sum of variable cost per unit (adjusted for learning) plus standard fixed cost per unit (adjusted for production rate). This model incorporates the two factors presumed to impact unit costs that have been most extensively investigated: cumulative quantity (Q) and production rate per period (PR).[6] It is consistent with both the theoretical and empirical literature which sees the primary impact of learning to be on variable costs and the primary impact of production rate to be on the spreading of fixed costs (Smith, 1980).

Simulated cost data in this study was generated using equation 9, while varying values for the variables and parameters on the right hand side of the equation to reflect differing conditions. In a strict sense, the results of this study are only applicable to situations where cost can be said to

[6]Smith (1980, 1981), for example, used a model similar to equation 9 to explore the effect of different production rates on unit cost. Balut (1981) and Balut, Gulledge and Womer (1989) construct models based on learning and production quantity to assist in "redistributing" overhead and "repricing" unit costs when changes in production rate occur. The Balut and Balut, Gulledge and Womer models differ in that they determine a learning rate for total (not variable) unit cost and then apply an adjustment factor to allow for the impact of varying production quantity on the amount of fixed cost included in total cost.

be generated from a function like equation 9. Unfortunately, "real world" cost functions are typically unknown. (If they were known, there would be no need to estimate approximations such as learning curves.) In practice the results of study are likely to be applicable to situations where cost comes from a process which has the same characteristics as those embodied in equation 9; specifically, a process in which some portion of cost is variable, and subject to learning effects, and some portion of cost is fixed, and assigned to units on the basis of production volume.

(3) The Simulated Conditions: The general research hypothesis is that the relative predictive accuracy of the LC and RA models will depend on the circumstances in which they are used. What conditions might be hypothesized to affect prediction accuracy? Seven different factors (independent variables) were varied during the simulation. In the following paragraphs, each factor is discussed. A label for each is provided, along with a discussion of why the factor may be relevant to prediction accuracy and how the factor was operationalized in the simulation. Table 5-1 summarizes the seven factors.

i) Data History (DATAHIST): The number of data points available to estimate parameters for a model should affect the accuracy of a model. More data available during the "estimation period" should be associated with greater accuracy for both the LC and the RA model.[7] But increased data may not impact the accuracy of the two models in the same manner. The RA model requires the estimation of an additional parameter, and consequently is likely to require more data points to avoid unreliable parameter estimates and resultant prediction inaccuracy. Hence it may be hypothesized that the RA model will be more sensitive (less accurate) when data is lean.

In the simulation, data history was varied from four to seven to ten data points available for estimating model parameters. This simulates having knowledge of costs and quantities for four, seven or ten production lots. Four is the minimum number of observations needed to estimate the parameters of the RA model by regression. The simulation focuses on lean data availability both because the effects of marginal changes in data availability should be most pronounced when few observations are available and because many real world applications (e.g., cost analysis of Department of Defense weapon system procurement) occur under lean data conditions.

---

[7]There are, of course, cost/benefit tradeoffs. The marginal benefits of increased prediction accuracy for any model must be weighted against the marginal costs of additional data collection.

Table 5–1 Independent Variables

CONCEPT	LABEL	LEVELS		
Data History	DATAHIST[1]	4	7	10
Variable Cost Learning Rate	LEARNRAT	75%	85%	95%
Fixed Cost Burden	BURDEN[2]	15%	33%	50%
Production Rate Trend	PROTREND[3]	Level		Growth
Production Rate Instability/Variance	RATEVAR[4]	0.05	0.15	0.25
Cost Noise/Variance	COSTVAR[5]	0.05	0.15	0.25
Future Production Level	FUTUPROD[6]	Low	Same	High

1 Number of data points available during the model estimation period; simulates the number of past production lots.

2 Standard per unit fixed cost as a percentage of cumulative average per unit total cost, during the model estimation period.

3 A level trend means production at 100% of standard production for each lot during the estimation period. A growth trend means production rate gradually increasing to 100% of standard production during the estimation period. The specific growth pattern depends on the number of production lots in the estimation period, with sequences as follows (expressed as a % of standard): for DATAHIST = 4: 33%, 67%, 100%, 100%. For DATAHIST = 7: 20%, 40%, 60%, 80%, 100%, 100%, 100%. For DATAHIST = 10: 10%, 20%, 35%, 50%, 70%, 90%, 100%, 100%, 100%, 100%.

4 Coefficient of variation of production rate. (Degree of instability of production rate around the general production rate trend.)

5 Coefficient of variation of total per unit cost.

6 "Same" means production rate at 100% of standard for each lot produced within the prediction zone. "Low" means production rate at 50%. "High" means production rate at 150%.

ii) Variable Cost Learning Rate (LEARNRAT): In the cost generating function, learning affects total unit cost by affecting variable cost per unit. Past research (Smunt, 1986) has shown that the improvement in prediction accuracy from including a learning parameter in a model (when compared to its absence) depends on the degree of learning that exists in the underlying phenomena being modeled; the greater the degree of learning, the more benefit gained from inclusion of a learning rate term. This suggests that prediction accuracy may depend on the learning rate and hence that relative prediction accuracy of the LC and RA models may also depend on the learning rate. In the simulation, variable cost learning rate (reflected in parameter d in equation 9) was varied from 75% to 85% to 95%. Generally, complex products or labor intensive processes tend to experience high rates of learning (70–80%) while simple products or machine-paced processes experience lower (90–100%) rates (Smunt, 1986).[8]

iii) Fixed Cost Burden (BURDEN): In theory (and in the cost function, equation 9) a change in the number of units produced during a period affects unit cost in two ways: First, increasing volume increases cumulative quantity and decreases variable cost per unit, due to learning. Second, increasing volume increase the production rate for a period and reduces fixed cost per unit, due to the spreading of total fixed cost over a larger output. Both these effects operate in the same direction; i.e., increasing volume leads to lower per unit cost. This has led some cost analysts to conclude that in practice, it is sufficient to use an LC model, letting the cumulative quantity variable reflect the dual impacts of increased volume. Adding a production rate term to an LC model is seen as empirically unnecessary.

In principle, if fixed cost was zero, cumulative quantity would be sufficient to explain total unit cost and production rate would be irrelevant.[9] But as fixed cost increases as a proportion of total cost, the impact of production rate should become important. This suggests that the relative predictive accuracy of the LC and RA models may depend on the amount of fixed cost burden assigned to total cost.

Fixed cost burden was simulated by varying the percentage of total unit cost made up of fixed cost.[10] Three percentages were used in the simulation: 15%, 33%, and 50%. The different percentages can be viewed as simulating different degrees of operating leverage, of capital

---

[8]See Conway and Schultz (1959) for further elaboration of factors impacting learning rates.

[9]This assumes that Eq. 9 is an adequate representation of the cost function. In equation 9 production rate only influences the assignment of fixed costs to units. Variable cost is unaffected by production rate. While there is some evidence that variable cost may be influenced by production rate in some situations, the dominant and unambiguous role of production rate appears to be in distributing fixed cost across units (Smith, 1980).

[10]Operationally this is a bit complex, since both per unit variable and per unit fixed cost depend on other simulation inputs (cumulative quantity and production rate per period). The process of relating fixed cost to total cost was as follows: First, a cumulative average per unit variable cost for all units produced during the estimation period was determined. Then a standard fixed cost per unit was set relative to the cumulative average per unit variable cost. For example, if standard fixed cost per unit was set equal to cumulative average variable cost per unit, then "on average" fixed cost would comprise 50% of total unit cost during the estimation period. Actual fixed cost per unit may differ from standard fixed cost per unit if the production rate (discussed later) was not at 100% of standard.

intensiveness, or of plant automation. The 15% level reflects the average fraction of price represented by fixed overhead in the aerospace industry, as estimated at one time by DOD (Balut, 1981).[11] The larger percentages are consistent with the trend toward increased automation (McCullough and Balut, 1986).

iv) Production Rate Trend (PROTREND): When initiating a new product, it is not uncommon for the production rate per period to start low and trend upward to some "normal" level. This may be due both to the need to develop demand for the output or the desire to start with a small production volume, allowing slack for working bugs out of the production process. Alternatively, when a "new" product results from a relatively small modification of an existing product, sufficient customer demand or sufficient confidence in the production process may be assumed and full scale production may be initiated rapidly. In short, two different patterns in production volume may be exhibited early on when introducing a new item: a gradual growing trend toward full scale production or a level trend due to introduction at full scale production volume.

How might the production rate trend during the estimation period affect the relative prediction accuracy of the LC and the RA models? Two contrary arguments seem relevant. On the one hand, the purpose of an RA model is to explain variance in cost due to variance in production rate. If production rate is virtually level during a model estimation period, there would be little period–to–period differences in unit cost caused by the spreading of fixed costs over varying outputs. Any rate effects would likely be swamped by other random impacts on cost. This suggests greater incremental benefit to using an RA model when the production rate has not been level, i.e., when growth occurred during the model estimation period. On the other hand, a growing production rate from period to period results in statistical problems. Within the RA model, there will be greater colinearity between cumulative quantity (necessarily growing each period) and production rate. Empirically, this colinearity has been observed by many cost analysts (e.g., Gulledge and Womer, 1986). The colinearity makes production rate a somewhat redundant variable and causes unreliable parameter estimates. This suggests less incremental benefit to using an RA model if production rate has been growing.

---

[11]In the absence of firm–specific cost data, the Cost Analysis Improvement Group in the Office of the Secretary of Defense treats 15% of the unit price of a defense system as representing fixed cost (Pilling, 1990).

Which argument holds is an empirical question. The simulation created two production trends during the model estimation period: "level" and "growth." These represented general trends (but, as will become clear momentarily, variance around the general trend was introduced). The level trend simulated a production rate set at a "standard" 100% each period during model estimation. The growth trend simulated production rate climbing gradually to 100%. Details of the trends are in Table 5–1.

v) Production Rate Instability/Variance (RATEVAR): Numerous factors, in addition to the general trend in output discussed above, may operate to cause period–to–period fluctuations in production rate. Manufacturers typically do not have complete control over either demand for output or supply of inputs. Conditions in either market can cause instability in production rate. (Of course, unstable demand, due to the uncertainties of annual budget negotiations, is claimed to be a major cause of cost growth during the acquisition of major weapon systems by the DoD). The concern here is with the likely effect of rate instability on the relative accuracy of LC and RA models. If production rate is highly stable across periods, there would be little variance in cost outcomes on which to estimate a rate parameter. Rate effects would be dominated by other impacts on cost. This suggests greater benefit to using an RA model when production rate is unstable over time.

Production rate instability was simulated by adding random variance to each period's production rate during the estimation period. The amount of variance ranged from a coefficient of variation of 0.05 to 0.15 to 0.25. For example, if the production trend was level and the coefficient of variation was 0.05 then "actual" production rates simulated were generated by a normal distribution with mean equal to the standard production rate (100%) and sigma equal to 5%.

vi) Cost Noise/Variance (COSTVAR): From period to period there will be unsystematic, unanticipated, non–recurring, random factors that will impact unit cost. Changes in the cost, type or availability of input resources, temporary increases or decreases in efficiency, and unplanned changes in the production process are all possible causes. Conceptually such unsystematic factors can be thought of as adding random noise to unit cost. While unsystematic variation in cost cannot (by definition) be controlled, it is often possible to characterize different production processes in terms of the degree of unsystematic variation; some processes are simply less well–understood, more uncertain, and less stable than others.

Does the relative predictive accuracy of LC and RA models depend on the stability of the process underlying cost? To investigate this question, random variance was added to the simulated costs generated from the cost function. The amount of variance ranged from a coefficient of variation of .05 to .15 to .25. For example, when the coefficient of variation was .25, then "actual" unit costs simulated were generated by a normal distribution with mean equal to cost from equation 9 and sigma equal to 25%.

vii) Future Production Level (FUTUPROD): Once a model is constructed (from data available during the estimation period), it is to be used to predict future cost. The production rate planned for the future may vary from past levels. Further growth may be planned. Cutbacks may be anticipated. Will the level of the future production rate affect the relative predictive accuracy of the LC and RA models? Is one model more accurate if cutbacks in production are anticipated and another if growth is planned? Conflicting scenarios can be argued. For example, on the one hand, inclusion of a rate term might be expected to increase prediction accuracy when production rate changes significantly (i. e., either growth or decline in the future period). On the other hand, if growth during the future prediction period merely continues a growth trend established during the estimation period, then production rate will once again tend to be correlated with cumulative quantity, and the incremental benefit of an RA model over an LC model may be less obvious.

In the simulation, future production was set at three levels: low (50% of standard), same (100% of standard) and high (150% of standard). These simulate conditions of cutting back, maintaining or increasing production relative to the level of production existing at the end of the model estimation period.

(4) The Cost Objective: What is to be predicted? Up to this point the stated purpose of the study has been to evaluate accuracy when predicting future cost. But which future cost? Three alternatives were examined.

i) Next period average unit cost: As the label suggests this is the average per unit cost of items in the production "lot" manufactured in the first period following the estimation period. Here the total cost of producing the output for the period is simply divided by the output volume to arrive at unit cost. Attention to this cost objective simulates the need to predict near term unit cost.

ii) Total cost over a finite production horizon: The objective here is to predict the total cost of all units produced during a fixed length production horizon. Three periods were used as the length of the production horizon (one production lot produced each period). If the future production rate is low (high) then relatively few (many) units will be produced during the finite production horizon. Attention to this cost objective simulates the need to predict costs over some specific planning period, regardless of the volume to be produced during that planning period.

iii) Total program cost: The objective here is to predict total cost for a specified number of units. If the future production rate is low (high) then relatively more (fewer) periods will be required to manufacture the desired output. The simulation was constructed such that at a low (same, high) level of future production six (three, two) future periods were required to produce the output. Attention to this cost objective simulates the need to predict total cost for a particular production program, regardless of the number of future periods necessary to complete the program.

Examining each of these three cost objectives was deemed necessary to provide a well-rounded investigation of predictive accuracy. However, the findings were substantially the same across the three cost objectives. In the interests of space, the remainder of this paper will discuss the analysis and results only for the first cost objective, the average cost per unit for the next period's output.

(5) The Measures of Prediction Performance: In order to compare prediction accuracy between LC and RA models, two kinds of measures were required. First, a model prediction error (ERR) was determined separately for each (LC or RA) model as follows:

$$ERR = | PUC - AUC | \div AUC$$

where

PUC = Predicted unit cost from either the learning curve or the rate adjustment model
AUC = Actual unit cost as generated by the cost function.

This is a commonly used error measure, the absolute percentage error.

Next a simple difference in errors (ERRDIFF) between the LC prediction and the RA prediction was calculated.

$$ERRDIFF = ERR_{LC} - ERR_{RA}$$

where

ERR_{LC}	=	Absolute percentage error using the learning curve model prediction of cost
ERR_{RA}	=	Absolute percentage error using the rate adjustment model prediction of cost.

Positive values for ERRDIFF mean that the rate adjustment model produced smaller prediction errors that the learning curve model and consequently imply increased accuracy from incorporating a production rate variable into the prediction process. Negative values mean the learning curve model is more accurate. ERRDIFF represents the dependent variable in the statistical analysis. The research question then becomes: What factors or conditions explain variance in ERRDIFF?

Figure 5–1 summarizes the complete simulation process leading up to the determination of ERRDIFF. The simulation was run once for each possible combination of treatments. Given seven factors varied and three possible values for each factor (except for PROTREND which had two), there were 3 x 3 x 3 x 3 x 3 x 3 x 2 = 1458 combinations. Thus the simulation generated 1458 observations and 1458 values for ERRDIFF.[12]

6. ANALYSIS AND FINDINGS

The simulation results were evaluated using analysis of variance (ANOVA) to conduct tests of statistical significance. All main effects are discussed in the following section. First order (pairwise) interactions were also tested. Those significant at .01 are also discussed. Table 6–1 provides the ANOVA results.

[12]In the simulation, just as in the real practice of cost analysis, it is possible for a model estimated on limited data to be very inaccurate, leading to extreme values for ERR (or ERRDIFF). If such outlier values were to be used in the subsequent analysis, findings would be driven by the outliers. Screening of the observations for outliers was necessary. During the simulation, if a model produced an ERR value in excess of 100%, then that value was replaced with 100%. This truncation has the effect of reducing the impact of an outlier on the analysis while still retaining the observation as one that exhibited poor accuracy. Alternative approaches to the outlier problem included deletion instead of truncation and use of a 50% ERR cutoff rather than the 100% cutoff. Findings were not sensitive to these alternatives.

Figure 5–1 Simulation Flowchart

Table 6–1 Analysis of Variance Results

SOURCE	DF	SUM OF SQUARES	MEAN SQUARE	F VALUE
Model	85	33.10921	0.3895	9.03
Error	1372	59.18006	0.0431	PR>F:
Corrected Total	1457	92.28927		0.0001
R^2		CV	ERRDIFF MEAN	
0.3588		1236.81	-.0168	

SOURCE	DF	ANOVA SS	F VALUE	PR≥F
DATAHIST	2	3.6780	42.63	0.000
LEARNRAT	2	.0510	0.59	0.5536
BURDEN	2	.7383	8.56	0.0002
PROTREND	1	1.6259	37.70	0.0001
RATEVAR	2	5.4728	63.44	0.0001
COSTVAR	2	9.3918	108.87	0.0001
FUTUPROD	2	0.6662	7.72	0.0005
DATAHIST*LEARNRAT	4	0.0699	0.41	0.8050
DATAHIST*BURDEN	4	0.1772	1.03	0.3920
DATAHIST*PROTREND	2	0.3217	3.73	0.0243
DATAHIST*RATEVAR	4	0.0844	0.49	0.7436
DATAHIST*COSTVAR	4	0.9813	5.69	0.0002
DATAHIST*FUTUPROD	4	0.7531	4.37	0.0016
LEARNRAT*BURDEN	4	0.2444	1.42	0.2260
LEARNRAT*PROTREND	2	0.0092	0.11	0.8979
LEARNRAT*RATEVAR	4	0.3933	2.28	0.0588
LEARNRAT*COSTVAR	4	0.2407	1.40	0.2332
LEARNRAT*FUTUPROD	4	0.3818	2.21	0.0655
BURDEN*PROTREND	2	0.0806	0.93	0.3931
BURDEN*RATEVAR	4	0.2325	1.35	0.2501
BURDEN*COSTVAR	4	0.7304	4.23	0.0021
BURDEN*FUTUPROD	4	1.0539	6.11	0.0001
PROTREND*RATEVAR	2	2.1867	25.35	0.0001
PROTREND*COSTVAR	2	0.1022	1.19	0.3059
PROTREND*FUTUPROD	2	0.4751	5.51	0.0041
RATEVAR*COSTVAR	4	0.3163	1.83	0.1199
RATEVAR*FUTUPROD	4	1.7734	10.28	0.0001
COSTVAR*FUTUPROD	4	0.8760	5.08	0.0005

Main Effects: As indicated in Table 6–1, all main effects, except LEARNRAT, are significant at .0005, indicating that values for ERRDIFF are significantly influenced by the treatment conditions. Table 6–2 summarizes ERRDIFF values under the various experimental conditions.

Table 6–2 Prediction Error Difference Between Learning Curve and Rate Adjustment Models: Main Effects

INDEPENDENT VARIABLE	ERRDIFF FOR EACH LEVEL		
DATAHIST Value:	4	7	10
ERRDIFF Mean:	–0.87	0.007	0.030
LEARNRAT Value:	75%	85%	95%
ERRDIFF Mean:	-0.009	-0.019	-0.023
BURDEN Value:	15%	33%	50%
ERRDIFF Mean	-0.047	-0.009	0.006
PROTREND Value	level	-	growth
ERRDIFF Mean	-0.050		0.017
RATEVAR Value:	0.05	0.15	0.25
ERRDIFF Mean:	-0.095	-0.010	0.054
COSTVAR Value:	0.05	0.15	0.25
ERRDIFF Mean:	0.082	-0.019	-0.114
FUTUPROD Value:	low	same	high
ERRDIFF Mean:	0.003	-0.007	-0.046

The major findings are perhaps made most obvious by a plot of ERRDIFF by treatments. Figure 6–1 shows a sample plot of ERRDIFF against DATAHIST. Two points are of interest. First, the relative accuracy of the LC and RA models clearly does depend on the number of observations available for model estimation. Second, both negative and positive values for ERRDIFF are present. This implies that under some conditions (few observations) the LC model is more accurate than the RA model. But under different conditions (larger number of observations) the RA model is more accurate. The crossover point occurs at about seven observations.

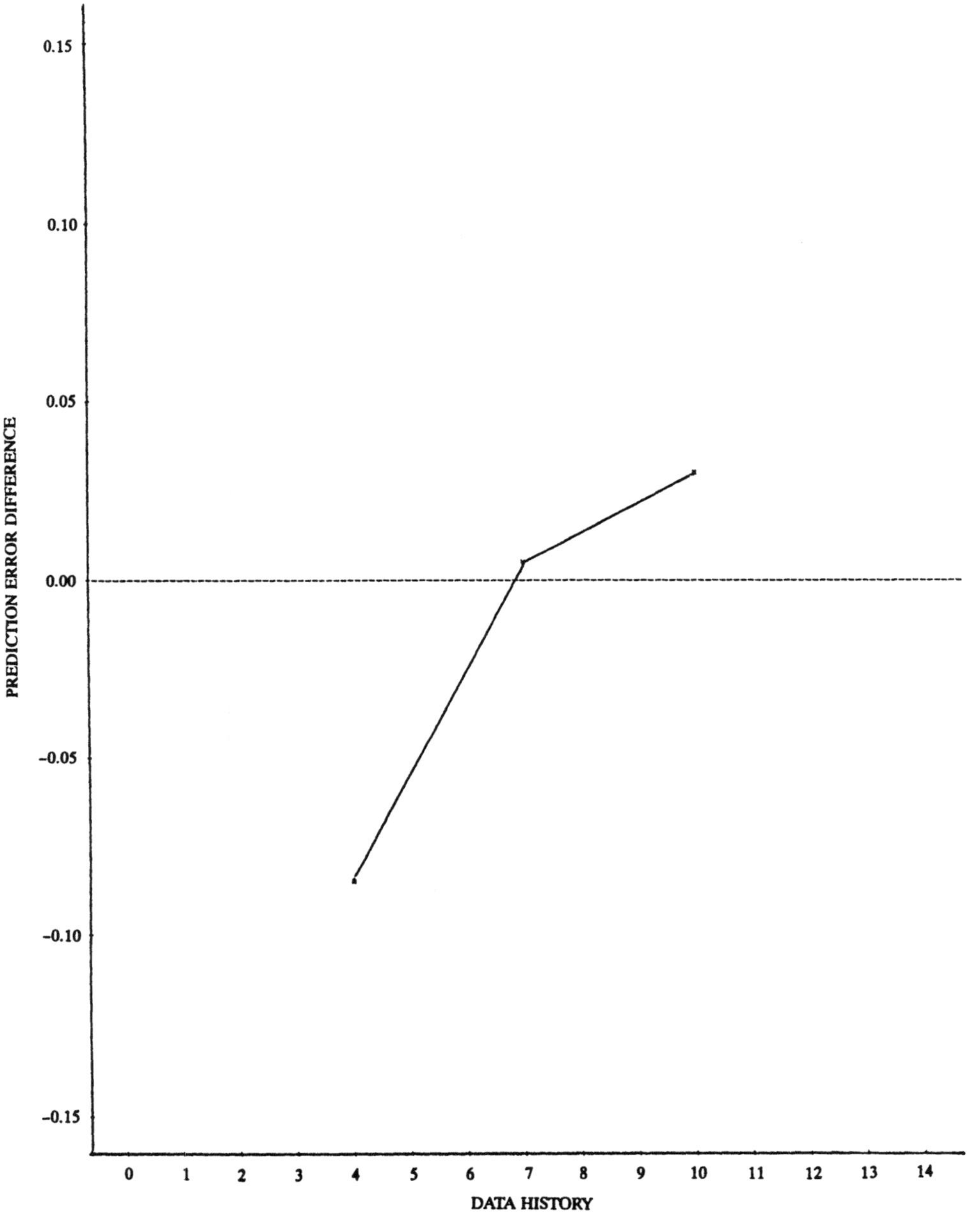

Figure 6–1 Plot of Prediction Error Difference by Data History

Figure 6–2 shows a similar plot of ERRDIFF with all (significant) variables superimposed. In this plot, 1, 2, and 3 on the x–axis reflect low, medium, and high values for the independent variables (which are taken from the left, middle and right columns of Table 6–2). Figure 6–2 makes several points. First for each of the six variables plotted, ERRDIFF varies significantly depending on the value (treatment) for the independent variable. This simply implies that the experimental conditions do reflect factors relevant to model accuracy.

Second for all six independent variables, at some point over the range of these variables, both positive and negative values for ERRDIFF exist. This means that neither the LC model nor the RA model dominates. Each model is more accurate under some experimental conditions.

Third, some factors apparently have a greater impact on relative prediction accuracy than others. For example, the plot shows that both the most extreme values for ERRDIFF (highest positive and lowest negative) are associated with different treatments on COSTVAR. The F–value for COSTVAR is also the highest and most significant in the ANOVA. This suggests that the greatest impact on relative predictive accuracy is due to variation in the amount of unsystematic noise impacting the measure of cost. A relatively large effect is also apparent due to variation in RATEVAR; while relatively small (but still significant) effects are associated with variations in BURDEN and FUTUPROD. These observations are interesting but must be interpreted with caution. COSTVAR does have the greatest impact on ERRDIFF over the particular range of treatments examined by the simulation. But if the ranges over which the individual independent variables were allowed to vary were altered, a different factor could appear to be most important.

Fourth, values for ERRDIFF are either monotonically increasing or decreasing for the various independent variable treatments. Hence some general conclusions about the impact of the experimental conditions on relative prediction errors are possible:

a) *Data History:* As indicated above the RA model outperforms the LC model when the number of observations for model estimation is relatively greater. This is expected; the benefit from greater precision due to the additional term in the RA model can only be realized when the data are sufficient to produce reliable parameter estimates.

b) *Fixed Cost Burden:* As expected the RA model tends to outperform the LC model as the proportion of total unit cost comprised of fixed cost increases. This implies that the benefit derived from use of a rate adjustment model depends on the capital intensiveness of the production process. Surprising, perhaps, was the fact that the change in ERRDIFF from the low (15%) fixed cost treatment to the high (50%) was relatively small.

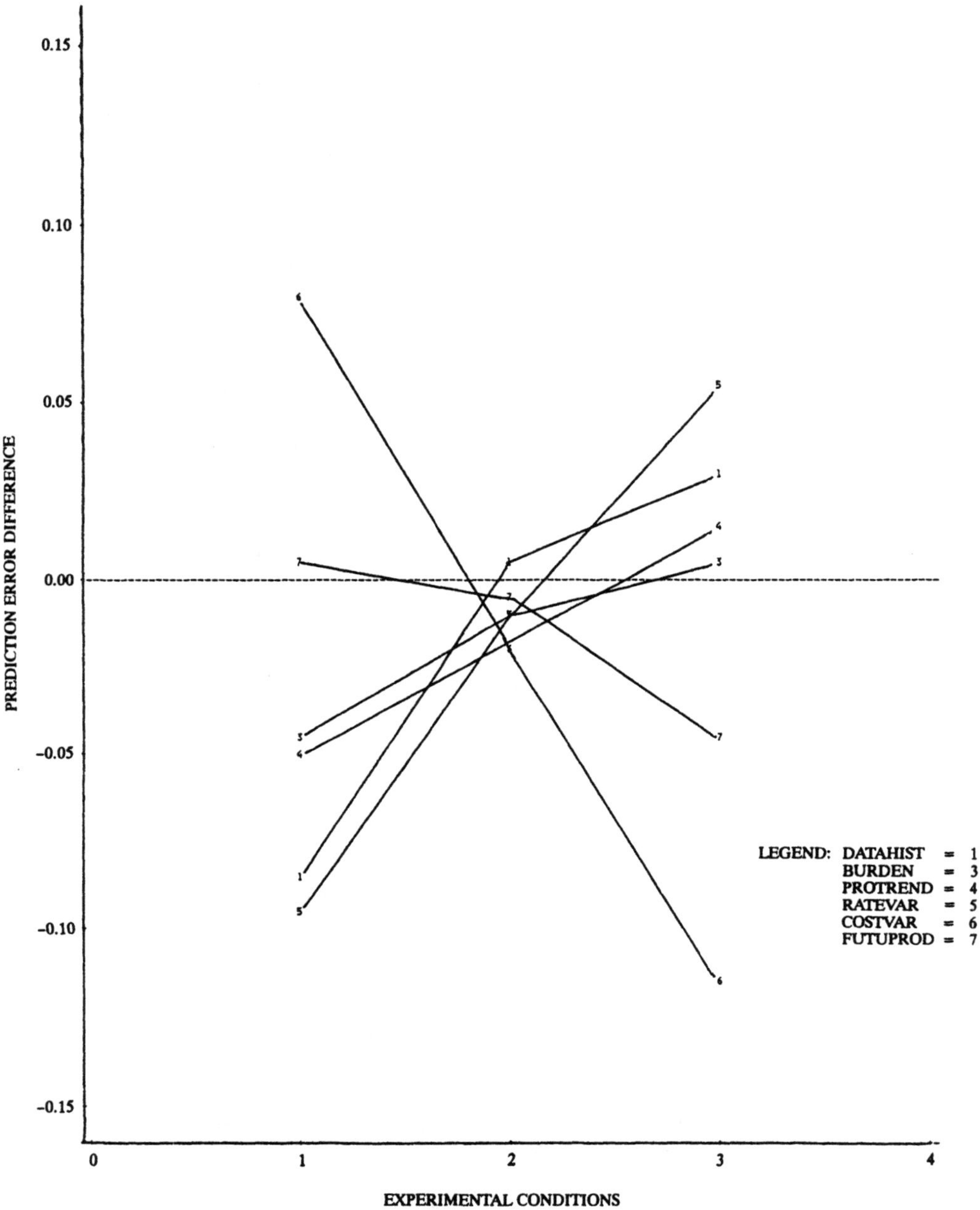

Figure 6–2 Plot of Prediction Error Difference by Main Effects

c) *Past Production Trend:* The RA model tends to outperform the LC model if the production rate was growing during the model estimation period. This finding suggests that it is necessary to have a changing production rate from period–to–period in order for the model estimation process to reliably capture the effect of rate on cost. It also suggests that relatively higher correlation between cumulative quantity and production rate (which will occur when rate has been growing) does not imply that cumulative quantity will be a sufficient variable. Cumulative quantity does not adequately capture the impact of rate changes.

d) *Production Rate Instability:* The RA model tends to outperform the LC model when there is greater instability in production rate. This suggests that it is desirable to have large period–to–period variation in production rate during the time of model estimation. In retrospect this conclusion seems obvious. If the variance in cost caused by rate changes is small, because the rate changes themselves are small, then the effect will be swamped by other effects and model parameters will be unreliable.

e) *Cost Noise/Variance:* As the random noise in unit cost increases, the RA model will perform more poorly relative to the LC model. This suggests that when there are unsystematic, unpredictable factors that influence cost, use of the RA model is less beneficial. The noise leads to unreliable estimates of the production rate effect and poorer cost predictions.

f) *Future Production Level:* The degree to which the RA model outperforms the LC model appears to be inversely associated with the production rate in the prediction period. When cutbacks in production are anticipated, cost is more accurately predicted by the RA model. When growth is anticipated, cost is more accurately predicted by the LC model. This suggests that the RA model is most beneficial when a reversing of the production trend from the estimation period to the prediction period is expected.

Interaction Effects: Eight first order interactions were also found to be significant in the ANOVA. Note that five of these interactions involve FUTUPROD as one of the interacting variables. This means that the impact of other variables on relative prediction accuracy is particularly sensitive to the anticipated level of production in the prediction period. Examining these eight interactions individually leads to further insight concerning the conditions that impact relative prediction accuracy. They are illustrated in Figs. 6–3 through 6–10.

Figure 6–3, the interaction of Data History with Future Production Level, demonstrates again that the RA model tends to outperform the LC model as the data history becomes richer. But this effect is most pronounced when a low future production level is planned. If a cutback in production is projected and few observations are available for model estimation, the LC model is substantially more accurate than the RA model (the most negative value for ERRDIFF occurs under these conditions). If a cutback in production is projected and relatively many observations are available for model estimation, the RA model is substantially more accurate than the LC model. (The most positive value for ERRDIFF occurs under these conditions.)

Figure 6–4, the interaction of Fixed Cost Burden and Future Production Level, demonstrates a somewhat analogous effect. The RA model tends toward better accuracy than the LC model as fixed cost burden increases. But again this effect is most pronounced when low levels of production in the future are planned. The analyst's concern for the impact of burden percentage on model accuracy should be greatest when volume cutbacks are expected.

Figure 6–5, the interaction of Production Rate Instability and Future Production Level also shows the importance of the future production level. The RA model tends to outperform the LC model as the degree of variability in the production rate increases. But the impact of rate variability is most pronounced when a cutback in production volume is planned. If the production rate was stable during the model estimation period, the LC model is more accurate. If the production rate was fluctuating and a cutback in volume is planned, then the RA model substantially outperforms the LC model.

Figure 6–6, the interaction of Cost Noise with Future Production Level, again illustrates how the future production volume impacts relative prediction accuracy. For all treatments, as unsystematic factors make cost more noisy, the RA model tends to lose its relative advantage over the simpler LC model. But this deterioration in predictive accuracy due to noise for the RA model is more evident when a cutback in future production volume is anticipated.

Figure 6–7 shows the interaction of the Production Trend during the estimation period and the Future Production Level during the prediction period. Again the steepest curve occurs when the future production level is low. This plot demonstrates that if production rate is relatively level during the model estimation period, the simple LC model tends to outperform the RA model. But the RA model outperforms the LC model when there is a shift or change in direction of the production trend from the estimation period to the prediction period. This follows from the positive value for ERRDIFF when a growing past production trend is coupled with a decline in future production level.

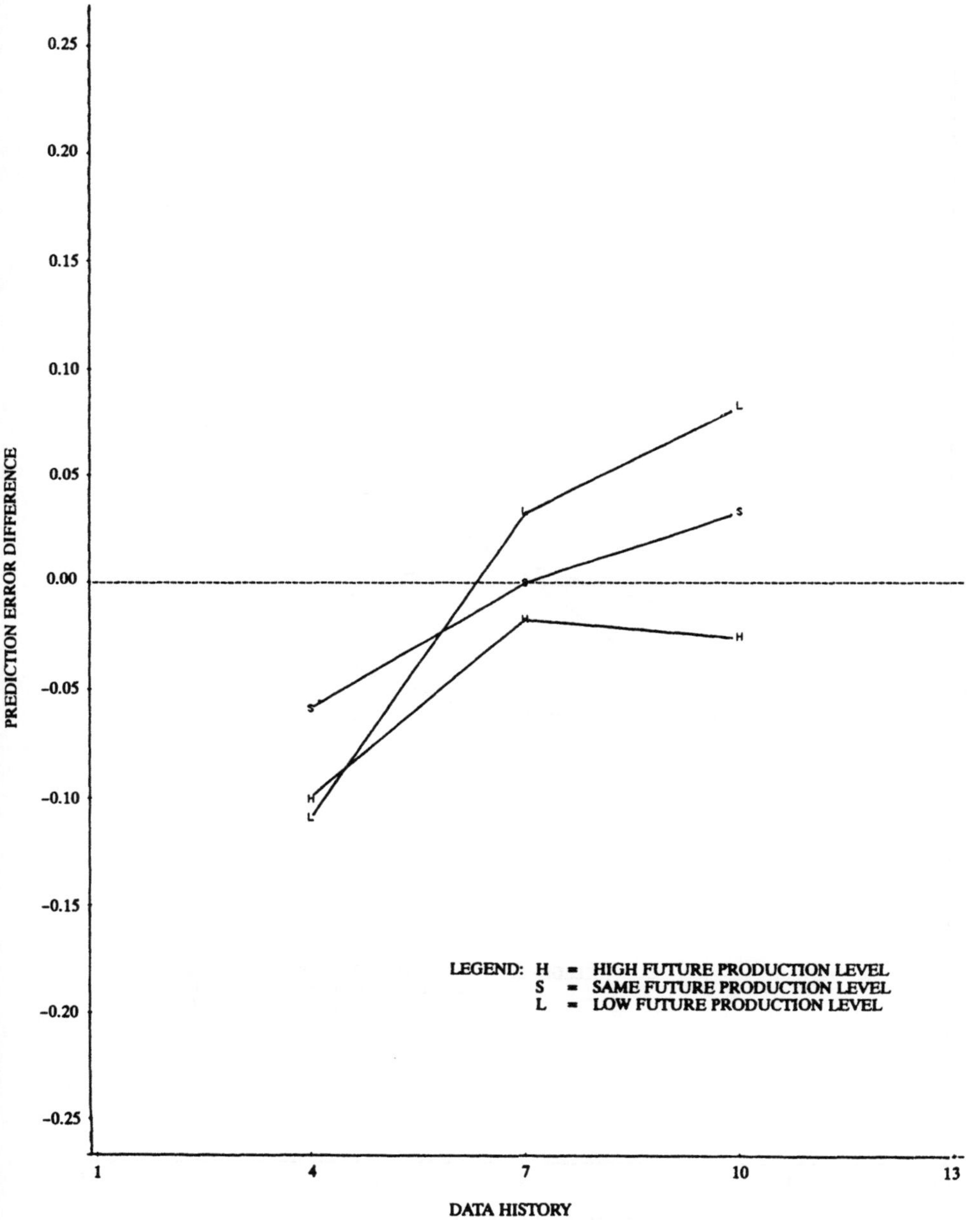

Figure 6–3 Interaction of Data History with Future Production Level

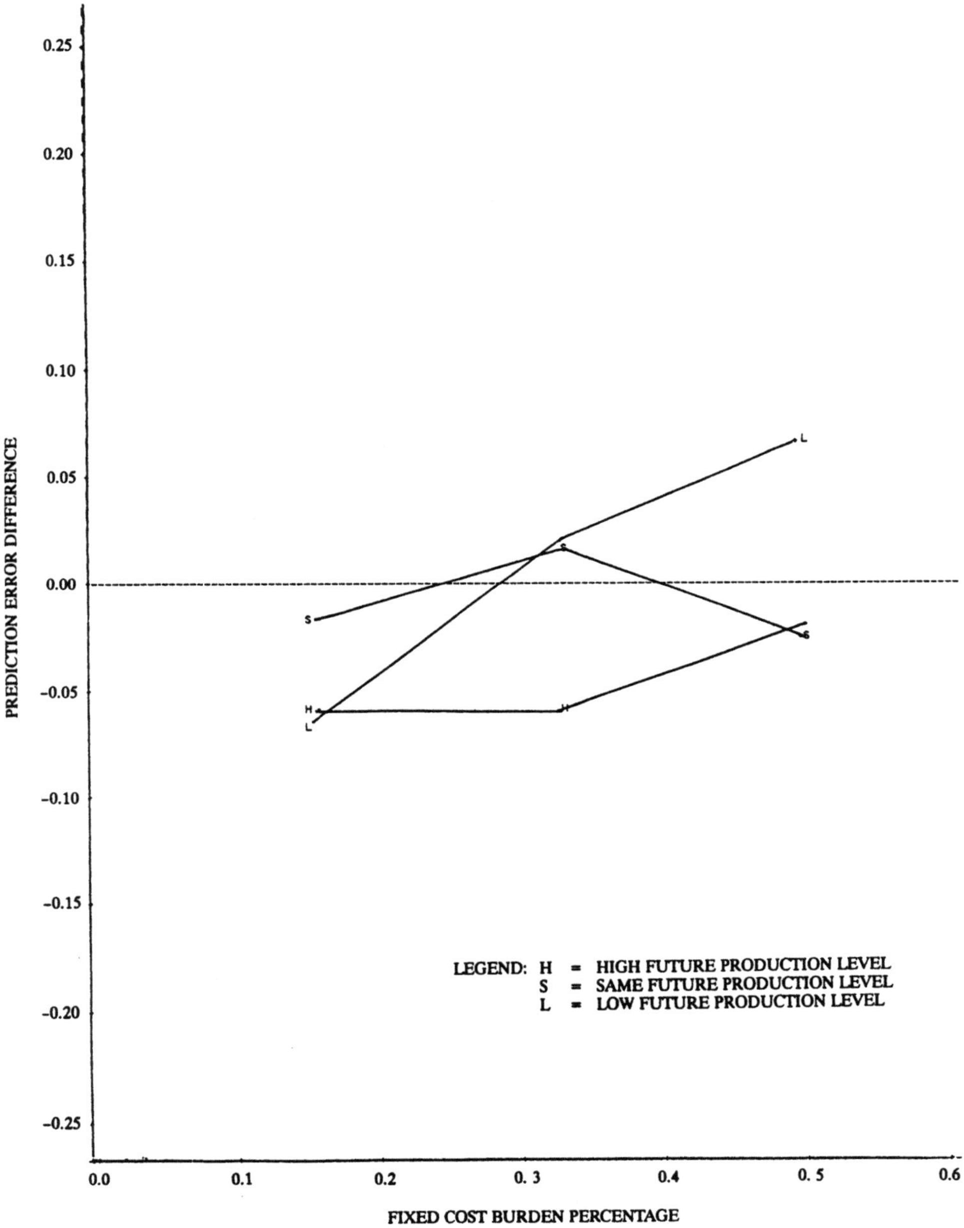

Figure 6–4 Interaction of Fixed Cost Burden with Future Production Level

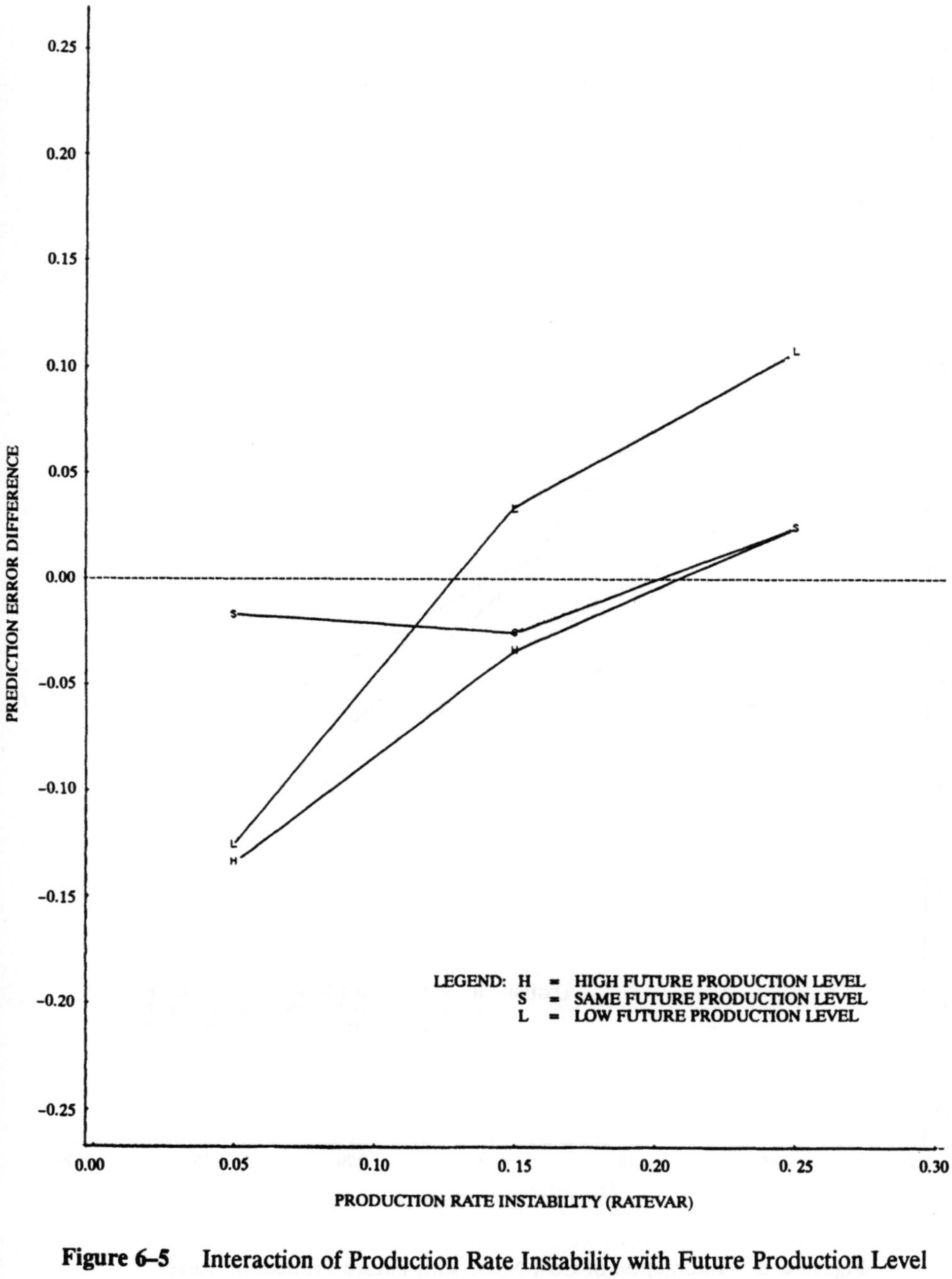

Figure 6–5 Interaction of Production Rate Instability with Future Production Level

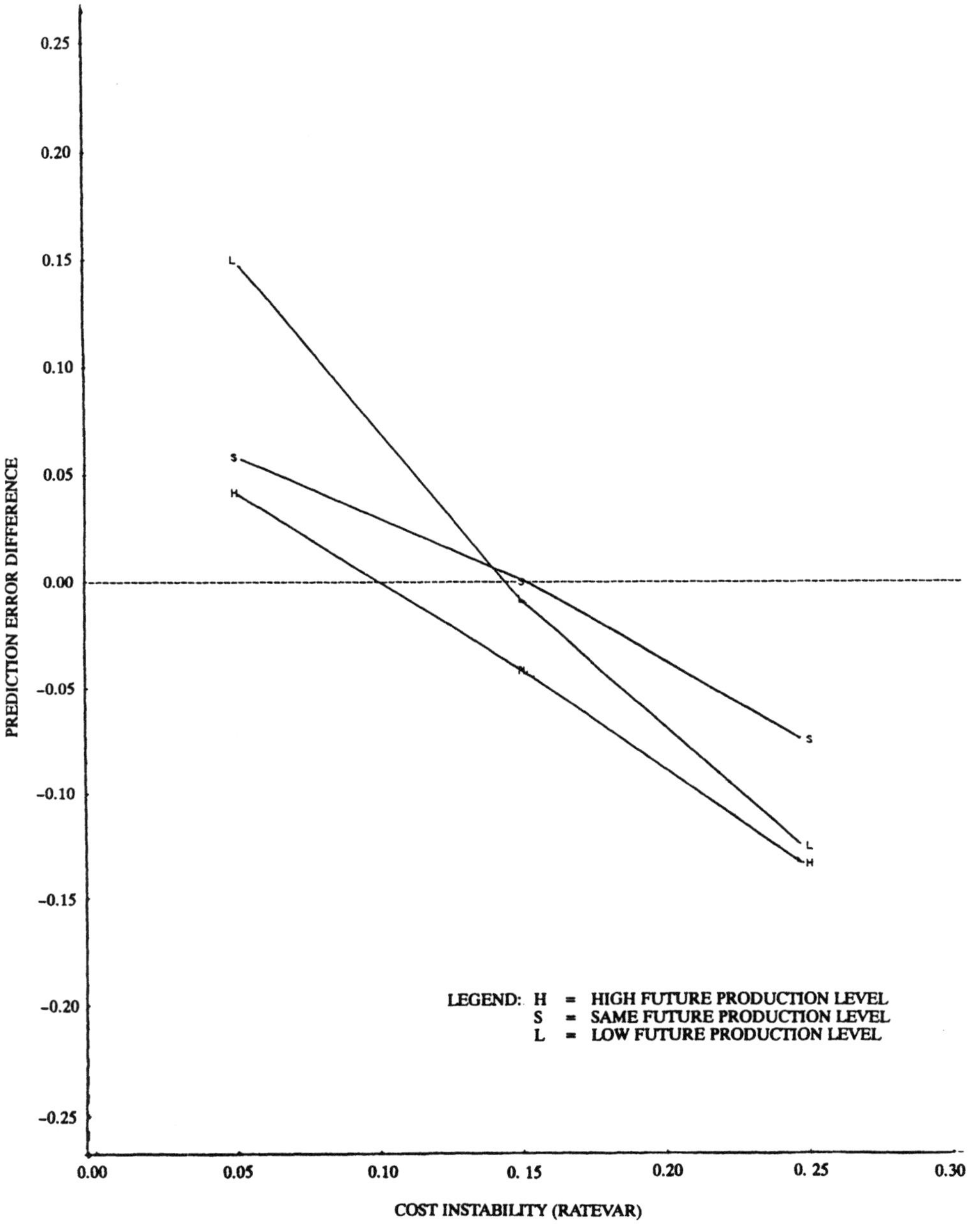

Figure 6–6 Interaction of Cost Noise with Future Production Level

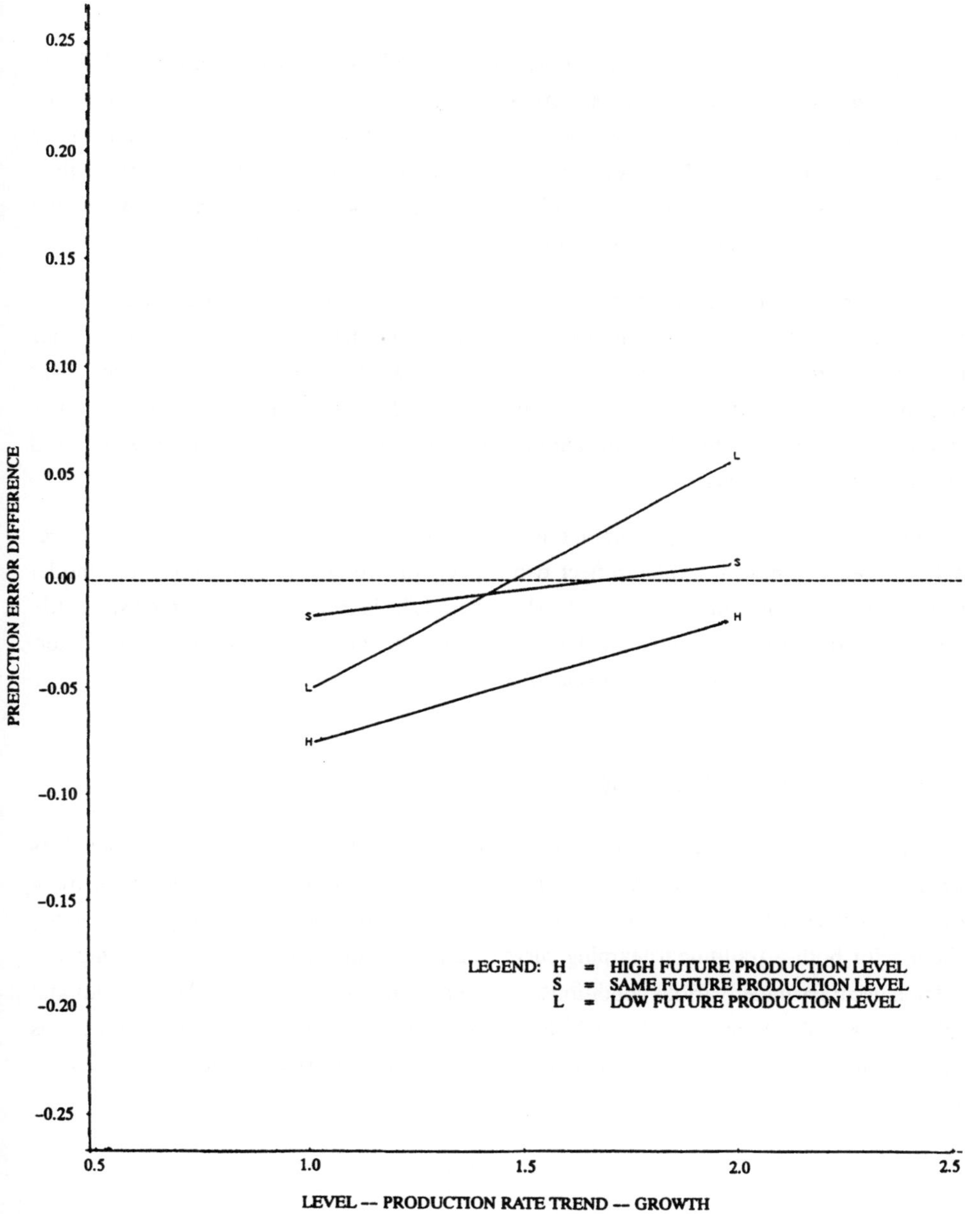

Figure 6–7 Interaction of Production Rate Trend with Future Production Level

Figure 6–8 shows an interesting interaction between Cost Noise and Fixed Cost Burden. The plot shows that under all conditions, increasing cost noise reduces, and then eliminates, any benefit from using the RA model. But the deterioration is most dramatic if the percentage of fixed cost in total cost is high. If burden is high and cost is not subject to unsystematic noise, then the RA model is substantially superior. But if burden is high and cost is subject to unsystematic noise, the RA model is substantially inferior.

Figure 6–9, the interaction between Cost Noise and Data History, illustrates a somewhat additive effect of two factors. Either increasing cost noise or reducing the number of observations causes deterioration in the performance of the RA model relative to the LC model. But combining the two effects magnifies the deterioration. If there is high cost noise and few observations on which to estimate model parameters, the RA model becomes very unreliable and the LC model is strongly superior.

Finally, Fig. 6–10 shows the interaction between Production Rate Instability and Production Rate Trend. These variables reflect two aspects of production rate during the model estimation period. The plot illustrates that when there is little trend in rate, coupled with little variability in rate relative to trend, then there is little basis on which to estimate a rate parameter. Hence the LC model strongly outperforms the RA model.

7. SUMMARY AND CONCLUSIONS

The central research question addressed in the study was: under what conditions does consideration of production rate and incorporation of a rate variable into a learning curve analysis of cost lead to a more accurate prediction of future cost? The analysis simulated prediction for both a traditional learning curve and a rate adjustment model and compared prediction accuracy under various conditions. One central finding was that neither a traditional learning curve analysis nor an analysis which considers the impact of production rate on cost is inherently superior. Neither model dominates; both outperform the other under some conditions.

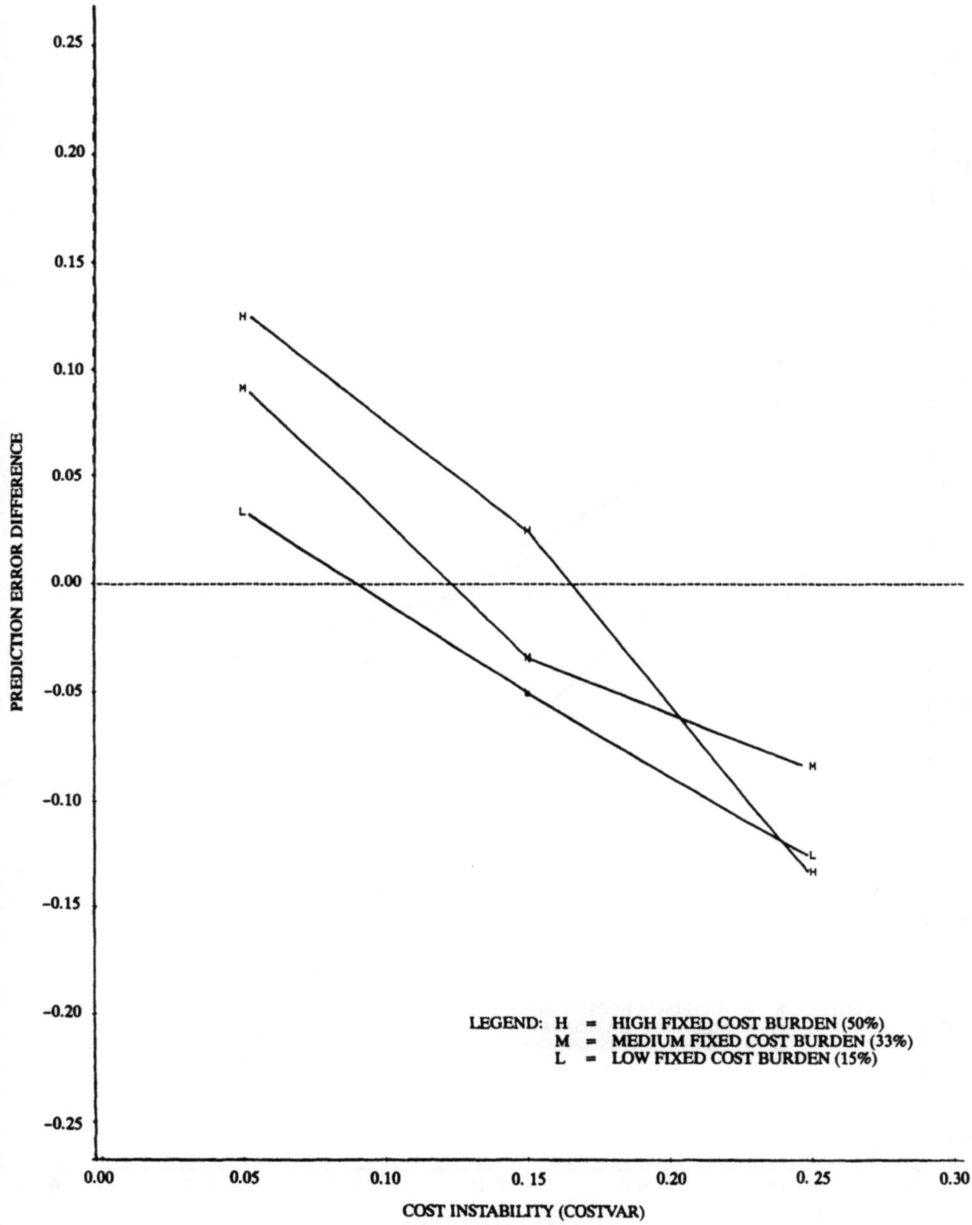

Figure 6–8 Interaction of Cost Noise with Fixed Cost Burden

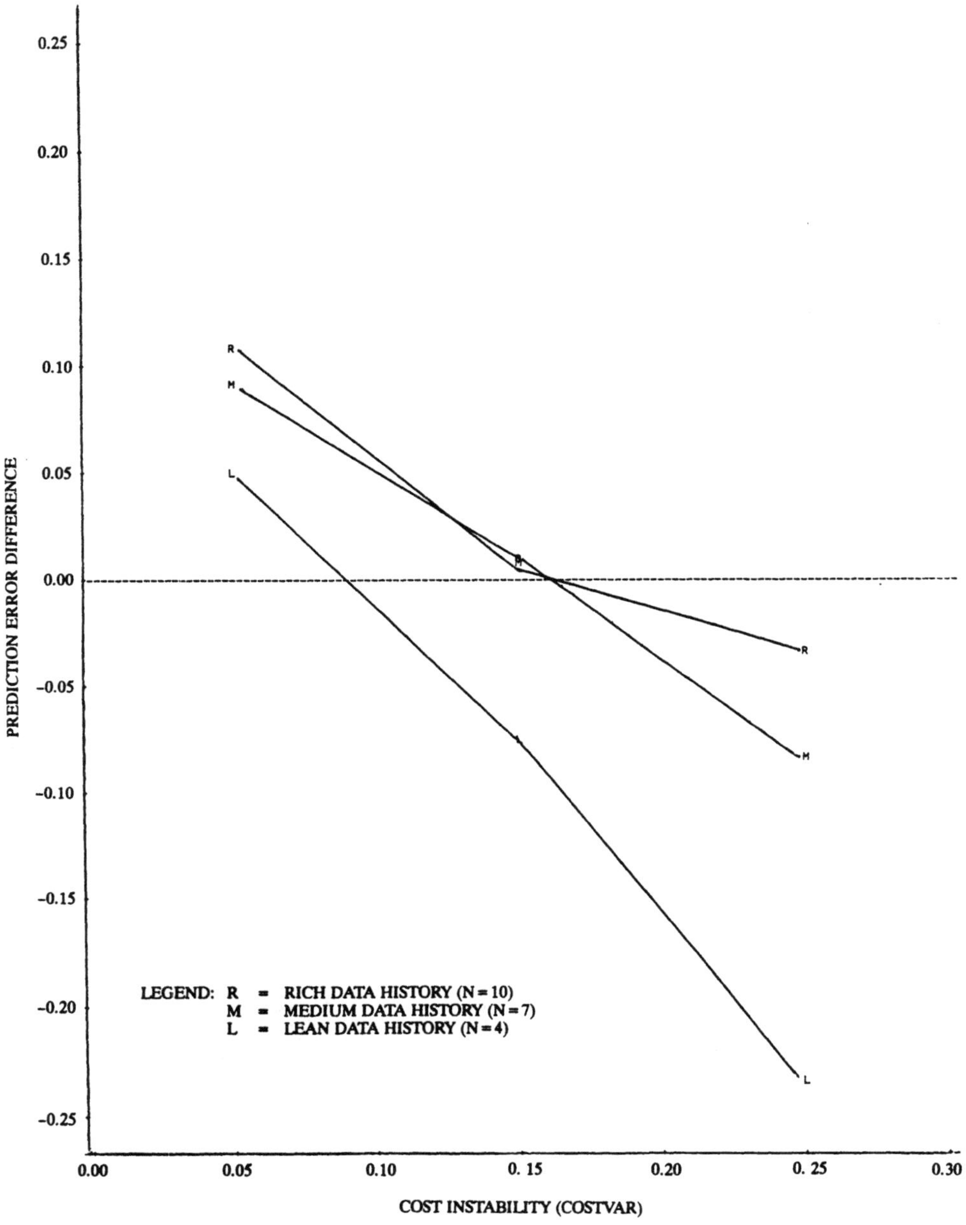

Figure 6–9 Interaction of Cost Noise with Data History

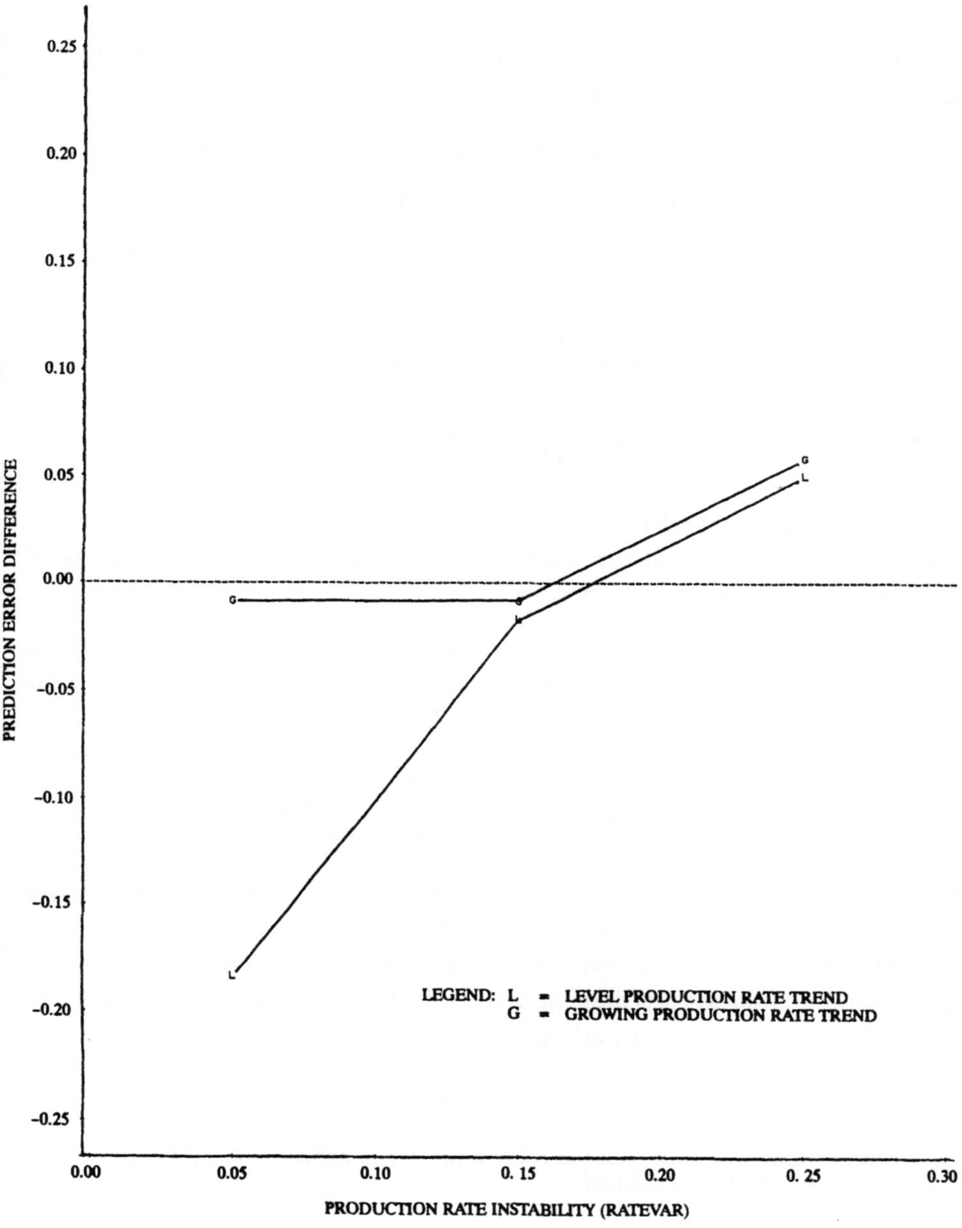

Figure 6–10 Interaction of Production Rate Instability with Production Rate Trend

General tendencies were evident. Considering production rate in the analysis lead to reduction in prediction error and improved accuracy when:

- The number of observations available for the analysis was relatively rich.
- The amount of fixed cost in total cost was relatively high.
- The production rate trend had been growing during the model estimation period.
- The period–to–period variability in production rate was relatively large.
- Random noise in cost due to unsystematic factors impacting cost was relatively low.
- Production volume was expected to be cutback in the future periods for which cost predictions were being made.

Each of these findings suggests that researchers or cost analysts, engaged in a cost prediction or cost analysis problem, may benefit from attending to such factors when deciding on the form of model or analysis they might bring to bear.

Numerous interacting impacts of combinations of factors on prediction accuracy were also evident, but a broad "theme" was apparent in the interactions, suggesting a general conclusion:

- The greatest impact (of changes in the various factors) on relative prediction accuracy (of the learning curve approach and the rate adjustment approach) occurs when cutbacks in future production are anticipated.

This means that researchers and cost analysts, attempting to predict future cost in an environment where future production volume is declining, will find the choice of an approach to be most critical. The relative accuracy of the learning curve approach or the rate adjustment approach is particularly sensitive to changes in data richness, fixed cost burden, production rate trend and stability, and cost noise when cutbacks are anticipated.

Other interactions collectively suggest a second general conclusion:

- The impacts of factors on the relative prediction accuracy of the two approaches tend to be additive.

This means that if the presence of one factor (say, lean data history) and the presence of another (say, high cost noise) both reduce the relative accuracy of the rate adjustment approach, then the presence of both will magnify the effect.

The conclusions of any study must be tempered by any limitations. The most prominent limitation of this study is the use of simulated data. Use of the simulation methodology was justified by the need to create a wide range of treatments and maintain control over extraneous influences. This limitation suggests some directions for future research.

- Re–analyze the research question while altering aspects of the simulation methodology. For example, are findings sensitive to the cost function assumed?
- Address the same research question using actual cost and production rate data. Are the same findings evident when using "real–world" data?

Providing confirmation of the findings by tests using alternative approaches would be beneficial.

Additional future research may be directed toward new, but related, research questions.

- Investigate other qualities of predictive performance, for example, bias. Do learning curves or rate adjustment models systematically provide predictions that are biased toward under or over estimation of future cost? Under what conditions?
- Investigate the magnitude of prediction errors using either a learning curve or rate adjustment approach. How large are average prediction errors under varying circumstances?
- Investigate other competing models or approaches to cost prediction. Perhaps accuracy can be improved by using some version of a "moving average" prediction model. Can such a model outperform both the learning curve and the rate adjustment approach? If so, under what circumstances?

REFERENCES

1. Alchian, A. (1963), "Reliability of Progress Curves in Airframe Production," *Econometrica*, Vol. 31, pp. 679–693.

2. Asher, H. (1956), *Cost–Quantity Relationships in the Airframe Industry*, R–291, RAND Corporation, Santa Monica, CA.

3. Balut, S. (1981), "Redistributing Fixed Overhead Costs," *Concepts*, Vol. 4, No. 2, pp. 63–72.

4. Balut, S., T. Gulledge, Jr., and N. Womer (1989), "A Method of Repricing Aircraft Procurement," *Operations Research*, Vol. 37, pp. 255–265.

5. Bemis, J. (1981), "A Model for Examining the Cost Implications of Production Rate," *Concepts*, Vol. 4, No. 2, pp. 84–94.

6. Boger, D. and S. Liao (1990), "The Effects of Different Production Rate Measures and Cost Structures on Rate Adjustment Models," in W. Greer and D. Nussbaum, editors, *Cost Analysis and Estimating Tools and Techniques*, Springer–Verlag, New York, 1990, pp. 82–98.

7. Cheney, W. (1977), *Strategic Implications of the Experience Curve Effect for Avionics Acquisition by the Department of Defense*, Ph. D. Dissertation, Purdue University, West Lafayette, IN.

8. Cochran, E. (1960), "New Concepts of the Learning Curve," *Journal of Industrial Engineering*," Vol. 11, pp. 317–327.

9. Conway, R. and A. Schultz (1959), "The Manufacturing Progress Function," *Journal of Industrial Engineering*, 10, pp. 39–53.

10. Cox, L. and J. Gansler (1981), "Evaluating the Impact of Quantity, Rate, and Competition," *Concepts*, Vol. 4, No. 4, pp. 29–53.

11. Dorsett, J. (1990), "The Impacts of Production Rate on Weapon System Cost," paper presented at the Joint Institute of Cost Analysis/National Estimating Society National Conference, Los Angeles, CA, June 20–22.

12. Gulledge, T. and N. Womer (1986), *The Economics of Made–to–Order Production*, Springer–Verlag, New York, NY.

13. Hirsch, W. (1952), "Manufacturing Progress Functions," *The Review of Economics and Statistics*, Vol. 34, pp. 143–155.

14. Large, J., H. Campbell and D. Cates (1976), *Parametric Equations for Estimating Aircraft Airframe Costs*, R–1693–1– PA&E, RAND Corporation, Santa Monica, CA.

15. Large, J., K. Hoffmayer, and F. Kontrovich (1974), *Production Rate and Production Cost*, R–1609–PA&E, The RAND Corporaton, Santa Monica, CA.

16. Levenson, G., et. al. (1971), *Cost Estimating Relationships for Aircraft Airframes*, R–761–PR, RAND Corporation, Santa Monica, CA.

17. Liao, S. (1988), "The Learning Curve: Wright's Model vs. Crawford's Model," *Issues in Accounting Education*, Vol. 3, No. 2, pp. 302–315.

18. Linder, K. and C. Wilbourn (1973), "The Effect of Production Rate on Recurring Missile Costs: A Theoretical Model," *Proceedings*, Eighth Annual Department of Defense Cost Research Symposium, Airlie VA, compiled by Office of the Comptroller of the Navy, 276–300.

19. McCullough, J. and S. Balut (1986), *Defense Contractor Indirect Costs: Trends, 1973–1982*, IDA P–1909, Institute for Defense Analysis, Alexandria, VA.

20. Moses, O. (1990), *Extensions to the Learning Curve: An Analysis of Factors Influencing Unit Cost of Weapon Systems*, Naval Postgraduate School Technical Report, NPS–54–90–016, Monterey, CA.

21. Pilling, D. (1989), *Competition in Defense Procurement*, The Brookings Institution, Washington DC, p. 35.

22. Smith, C. (1980), *Production Rate and Weapon System Cost: Research Review, Case Studies, and Planning Model*, APR080–05, U. S. Army Logistics Management Center, Fort Lee, VA.

23. Smith, C. (1981), "Effect of Production Rate on Weapon System Cost," *Concepts*, Vol. 4, No. 2, pp. 77–83.

24. Smith, L. (1976), *An Investigation of Changes in Direct Labor Requirements Resulting From Changes in Airframe Production Rate*, Ph. D. dissertation, University of Oregon, Eugene, OR.

25. Smunt, T. (1986), "A Comparison of Learning Curve Analysis and Moving Average Ratio Analysis for Detailed Operational Planning," *Decision Sciences*, Vol. 17, No. 4, Fall, pp. 475– 494.

26. Washburn, A. (1972), "The Effects of Discounting Profits in the Presence of Learning in the Optimization of Production Rates," *AIIE Transactions*, 4, pp. 255–313.

27. Wetherill, G. (1986), *Regression Analysis with Applications*, Chapman and Hall, New York.

28. Womer, N. (1979), "Learning Curves, Production Rate and Program Costs," *Management Science*, Vol. 25, No. 4, April, pp. 312–319.

29. Wright, T. (1936), "Factors Affecting the Cost of Airplanes," *Journal of Aeronautical Sciences*, Vol. 3, pp. 122–128.

30. Yelle, L. (1979), "The Learning Curve: Historical Review and Comprehensive Survey," *Decision Sciences*, Vol. 10, No. 2, April, pp. 302–328.

UNCERTAINTY AND CAPITAL INVESTMENT DECISIONS

Thomas A. Noll
Hercules Aerospace
P.O. Box 98, Magna, Utah 84044

ABSTRACT

Uncertainty is often incorporated into capital investment decisions by adjusting the discount rate. Adjusting the discount rate is a useful technique for sensitivity analysis, but discount rate adjustments may not adequately address the relative uncertainty of projects. This paper examines an alternative strategy which is based on probability theory and the analytical framework of the Capital Asset Pricing Model (CAPM). The advantage of the Capital Asset Pricing Model is that the variability in the return stream and the timing of the return stream are directly considered in the analysis. The results of the CAPM analysis can then be used to compare alternative projects and strategies.

1. INTRODUCTION

The problem of uncertainty has long complicated investment decisions, and although there is general agreement on the method of dealing with the problem, there is also general agreement on the inadequacy of the accepted techniques. The standard procedure is to increase the discount rate associated with the project as the uncertainty associated with the project increases. A variety of methods, both linear and nonlinear, have been proposed for the transformation of the discount rate, but all of the techniques are nearly identical in their results.

A higher discount rate generally favors projects with short-term planning horizons because the perceived risk exposure is lower in a shorter planning horizon. More specifically, a higher discount rate favors projects with up-front benefits, and long-term costs. Often the different projects are compared by contrasting the present value of the net income stream. The present value for a stream of income is calculated in discrete time:

$$PV = \frac{R_1}{(1+r)^1} + \frac{R_2}{(1+r)^2} + \ldots + \frac{R_N}{(1+r)^N} \tag{1}$$

where:

PV	=	discounted value of the income stream
r	=	discount rate
R	=	per-period income values

The above formula can be written in sigma notation:

$$PV = \sum_{n=1}^{N} \frac{R_n}{(1+r)^n}$$

where N equals the total number of periods in the income stream accounting, and n is equal to an individual element in the income stream.

The difficulty in applying the formula arises because increasing the discount rate reduces the present value of the outer period values. For example, consider a single period income return of $1,000 in period 10 ($n = 10$) for two different discount rates. When the discount rate equals 5% ($r = 0.05$), the discounted value of the $1,000 becomes $613. When the discount rate is raised to 10%, the discounted value of $1,000 in period 10 becomes $386. A plot showing the present value of $1,000 received in period 10 as a function of the discount rate is shown in Fig. 1-1.

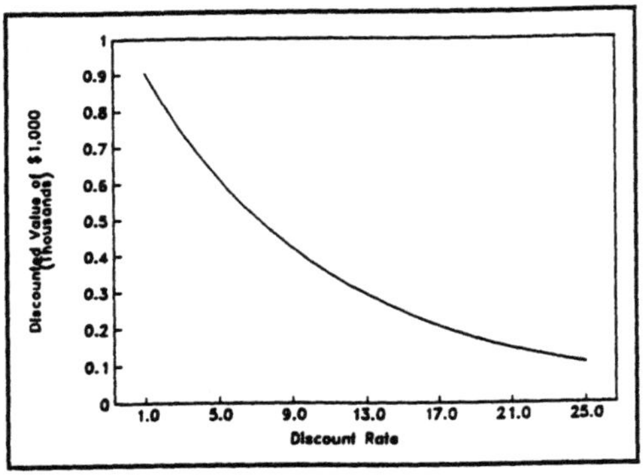

Figure 1-1 Discounted Value of $1,000 at 10 periods

Raising the discount rate favors projects with short–term returns and long–term costs; the investment analysis becomes myopic. Increasing the discount rate in response to uncertainty shifts the focus and the preference to projects with shorter planning horizons. The techniques used to increase the discount rate, linear transformations, nonlinear transformations, or haphazard transformations, all generate the same net result because all of the techniques raise the discount rate and the exact technique is irrelevant to the analysis; the net result is to favor short–term benefits and long–term costs. It would be possible for a short–term project to be accepted over a long–term project, even though the long–term project may have less uncertainty associated with its income stream.

Tables 1–1 and 1–2 show balance sheets for two sample projects. Both projects have a life of five periods, return an income stream of $200 in each period, and cost $400.

Table 1–1 Project 1 Balance Sheet

Period	1	2	3	4	5	Total
Project 1: Undiscounted Values						
Income	$200	$200	$200	$200	$200	$1,000
Probability	0.5	0.5	0.5	0.5	0.5	
Probable Income	$100	$100	$100	$100	$100	$500
Outflow					($400)	($400)
Net Project Benefit						**$100**
Project 1: Discounted Values Discount Rate = 10.0%						
Income	$182	$165	$150	$137	$124	$758
Probability	0.5	0.5	0.5	0.5	0.5	
Probable Income	$91	$83	$75	$68	$62	$379
Outflow					($248)	($248)
Net Project Benefit						**$131**

The per–period probability (risk) in project 1 is 50 percent, meaning that there is a 50 percent chance that the return in each period will actually be realized. This is a binary situation; the return is either $200 or zero. The probable income stream represents the expected income adjusted by the probability of return. The costs in project one are incurred during the fifth period. The bottom portion of Table 1–1 shows the discounted values of the income and costs associated

Table 1–2 Project 2 Balance Sheet

Period	1	2	3	4	5	Total
Project 1: Undiscounted Values						
Income	$200	$200	$200	$200	$200	$1,000
Probability	0.6	0.6	0.6	0.6	0.6	
Probable Income	$120	$120	$120	$120	$120	$600
Outflow	($400)					($400)
Net Project Benefit						**$200**
Project 1: Discounted Values Discount Rate = 10.0%						
Income	$182	$165	$150	$137	$124	$758
Probability	0.6	0.6	0.6	0.6	0.6	
Probable Income	$109	$99	$90	$82	$75	$455
Outflow	($364)					($364)
Net Project Benefit						**$91**

with project 1. The discount rate is set at 10 percent. The net return in project 1 is $100 in undiscounted dollars, and $131 in discounted dollars.

The balance sheet for project 2 is presented in Table 2–1. Project 2 is similar to project 1. The income stream and the cost of the two projects are equivalent. However, the probability of a positive return is greater in project 2 (60 percent in project 2 instead of 50 percent in project 1) and the cost in project 2 is incurred during the first period. The undiscounted value of project 2, $200, is greater than project 1. The discounted value of project 2 is shown in the bottom portion of Table II. Project 2 has a lower discounted value at $91, even with the increased probability of return. Under most methods of analysis, project 1 would be favored over project 2 because of the greater discounted net benefits.

It is also important to notice that although the discount analysis considers the timing of the income stream (projects with long–term costs are favored over up–front costs) the compatibility of the income stream timing with the projected cash flow of the firm is not considered. For a firm operating in a cyclical environment, the timing of income could be critical. Firms may consider projects with dampened cyclical income streams to be favorable, yet the discount analysis cannot incorporate the cyclical characteristics into the evaluation.

The theoretical framework of the Capital Asset Pricing Model (CAPM), frequently used in portfolio analysis, is well suited to the analysis of capital projects. The CAPM is based on probability theory, and can be adapted to accommodate the issues of uncertainty and income stream timing. Using the CAPM, projects with different income streams, up–front and long–term, and different levels of uncertainty can be equally compared.

2. THE CAPITAL ASSET PRICING MODEL

The CAPM is a finance tool that is commonly used to evaluate different investments and investment portfolios. The model is discussed in finance textbooks such as Garbade (1982) and Fama (1976), and is usually applied to stock market analysis. In the standard application, the expected return stream from an individual asset is evaluated in comparison with the overall return stream of a portfolio of assets. In this instance the return stream associated with the individual asset and the portfolio are not known with certainty, and the goal of the analysis is to determine the investment characteristics of the individual asset. Realistically, this scenario is no different than the situation a firm faces in evaluating a capital project; the return stream of the firm, and the return stream associated with the capital project are not known with certainty, and the goal is to determine the investment characteristics associated with the individual capital project.

The assumptions behind the strict derivation of the CAPM are that:

1. The returns associated with an asset are drawn independently from a single normal distribution.
2. Investors possess homogeneous expectations about an asset's performance. This means that all investors have equal access to the past history of an asset's return stream.
3. Investors preference functions are continuously differentiable. A higher return is preferred to a lower return, and a lower variance (risk) is preferred to a higher variance.

The CAPM treats the return stream as independent events drawn from a normal probability distribution. The return stream associated with an individual asset can be described with the normal probability distribution. The return stream is completely characterized by its mean and variance:

$$\text{Return Stream of Asset } i \sim N(\mu_i, \sigma_i^2) \tag{3}$$

The expected returns from asset *i* are its mean μ, and the uncertainty of the returns is described by the variance in its return stream σ_i^2. For example, consider a 5–year certificate of deposit. The expected returns from the asset might be 8%, or 1.08, and realistically, there is no uncertainty associated with the returns, so the variance is essentially zero. Contrast the Certificate of Deposit (CD) with the returns on a state–sponsored lottery: most lotteries return 50 cents on the dollar as prize money, leading to an expected return, $\mu_{Lottery}$, of 0.50, but there is considerable variability associated with the returns. Many players recover nothing from their investment, while a small group may experience returns of many thousands of dollars on a $1 ticket; a very favorable rate of return. The return on the lottery can be described with the bell–shaped probability curve associated with the normal distribution. The return stream on the certificate of deposit is also described by a normal distribution, but because the variance of the return is so small, the probability distribution appears as a single spike.

Even more important to investment analysis than an asset's variance is the covariance of the asset's return stream with the return stream of the market in general. The covariance measures the relationship between the changes in two variables. For example; if increases in one variable are always associated with increases in another, then the two variables would have a positive covariance. The covariance between two variables x and y is defined:

$$\text{Covariance of } x,y = \sigma_{x,y} = \frac{\sum_{i=1}^{N} (x_i - \mu_x)(y_i - \mu_y)}{N - 1} \tag{4}$$

where x_i and y_i are the ith values of variables x and y, and μ_x and μ_y are the mean values of x and y. N is the number of paired observations in the data set.

The concept of the covariance between an individual asset's returns and the returns of the market in general is central to the theoretical development of the CAPM. The covariance is used to quantify the risk associated with an asset. In the example above, the certificate of deposit would have a zero covariance with the market because there is no variation in the certificate of deposit return stream. The CD pays a constant rate of return regardless of what happens to the capital market in general. The covariance between a lottery ticket and the market would also be zero because a lottery ticket pays a rate of return that is independent of the capital market.

Formally, the CAPM is defined in the following relationship:

$$\mu_i - R_f = \beta_i \left(\mu_{Market} - R_f\right) \tag{5}$$

where:

R_f = the return to a risk-free asset such as a CD or a government security, and

$$\beta_i = \frac{\sigma^2_{i,Market}}{\sigma^2_{Market}} \tag{6}$$

The β measures the ratio of the marginal risk associated with security i to the risk of the market in general. The risk is defined as the variability in the asset stream, and is measured by the statistical concepts of variance and covariance discussed above. The statistical concepts are discussed more completely in Larson (1982).

The CAPM equation shows that the premium return on a security, that return in excess of the risk free return ($\mu_i - R_f$), is linearly proportional to the risk on that security, and the constant of proportionality is the β of the asset. A β of zero implies that there is no relation (the covariance is equal to zero) between the return of an asset and the returns in the market. Both the lottery ticket and the CD would have a β of zero. As the β of a security increases, so does the risk of earning the expected return on that security. A β equal to 1 indicates that the risk on that security is exactly equal to the average market risk; a β greater than 1 indicates that the asset returns are more risky than the market in general, and a β less than 1 indicates that the return stream associated with this asset is less risky than the market in general. A β equal to 1 indicates that the returns on that individual asset are exactly correlated with the returns on the market in general. There is no reason to believe that the β of an asset remains constant through time.

Much of the discussion surrounding the CAPM has been related to the measurement of the return stream of the market in general, measurement of the market portfolio. The market return stream has been difficult to quantify. For the evaluation of capital projects the problem is not so great. What is needed in project analysis is a benchmark against which the different alternative projects can be compared.

3. APPLYING THE CAPM TO PROJECT ANALYSIS

To apply the CAPM to project analysis, the values for several variables must be identified. These include:

R_f	=	the risk-free rate of return
μ_{Market}	=	the average rate of return for the market portfolio of investment alternatives
μ_i	=	the average rate of return for the investment under consideration
σ_{Market}	=	the variance of the return to the market portfolio
$\sigma_{i,Market}$	=	the covariance between the return on the market portfolio and the investment under consideration.

The risk-free rate of return is often specified to be the return on an asset such as U.S. Treasury notes. U.S. Treasury notes represent the income stream that could be earned by following a safe investment strategy.

The average rates of return for the individual asset i and the market are usually not too difficult to calculate. In most investment decisions the rates of return are known. To perform the traditional analysis and estimate the present value of the income stream, the per-period values of the income stream must be defined. The estimates can also be used to calculate the average rate of return in the CAPM analysis.

The variance of an income stream can also be calculated. When considering an investment that has a historical income stream, the mean and variance can be estimated from the historical data. One example of this situation is a company that is considering the acquisition of another firm or division. The mean and variance of the acquisition's income stream could be estimated based on the historical data.

Many possible investments do not have an associated history and for these investments the mean and variance must be approximated. This category includes most public and private sector investments. Although the exact income and cost streams are unknown, the CAPM structure can accommodate this uncertainty in either public sector projects (Wilson, 1982) or in private sector investment projects.

The estimated mean of the income stream should be known; the mean is the expected value of the income stream. The variance can be approximated, and the variance may entail a degree of subjective judgement. However, adjusting the discount rate in a traditional analysis is also subjective, and just as sensitivity analyses are conducted with different discount rates, similar sensitivity analyses can be performed with alternative estimates of the mean and variance.

In a standard normal distribution 95% of the area under the distribution is contained within two standard deviations from the mean; over 99% of the area falls within three standard deviations from the mean. Many analysts who cannot estimate the variance of the income stream will be able to discuss probability values for the income stream. For example the evaluation presented by an analyst might be: "The average rate of return for the investment is calculated at 10% and there is a 95% probability that the income stream will be within ±5 percentage points of the mean." The analyst has just identified the two–sigma boundary for the income stream, 5% to 15% rate of return. The two sigma boundary for the investment is equal to 5 percentage points, and one sigma equals 2.5 percentage points. The variance is equal to 2.52^2 or 6.25.

In many situations it might be easier to specify the variance on a per–period basis; the analyst may indicate the rate of return confidence boundaries for each of the years under consideration. Normal distributions are stable under addition. Adding some number of normal distributions results in another normal distribution but with different locational parameters, mean and variance. The mean of the return stream can be determined by averaging as mentioned earlier, and the variance is the sum of the individual variances.

The key assumption in calculating the mean and variance from a series of individual estimates is that the estimate of the investment performance is independent of the estimate of the performance in all other periods. This assumption may be somewhat unrealistic (a poor performance in one period may likely be followed by a poor performance in another period), but the assumption is certainly valid when using estimates of the income stream. The error in one performance estimate is unrelated to past estimating errors.

The formulas for determining the mean and variance of the income stream, where the income stream is composed of a number of individual estimates, are as follows:

$$\text{Mean} = \mu_{\text{Total}} = \frac{\sum_{t=1}^{T} a_t \mu_t}{T} \tag{7}$$

$$\text{Variance} = \sigma^2_{\text{Total}} = \frac{\sum_{t=1}^{T} a_t(\mu_t - \mu_{\text{Total}})^2}{T - 1} \tag{8}$$

The a's in the above expressions are weighting factors. If all of the periods are treated equally, then a_t equals 1. μ_t is the mean of the individual estimate; if μ_t is a combination of equally–weighted estimates, then a would equal one for each estimate, and μ would be that estimate value. Using different weighting schemes it would be possible to assign certain periods, or certain estimates, greater importance in the analysis. For most analyses, equal weighting of the different periods in the mean and variance formulas would be the appropriate choice. In the variance formula (equation 8), μ_{Total} refers to the quantity calculated in equation 7.

The values for the market portfolio of investment alternatives can be calculated in a similar fashion. One recommendation is to use the in–house sales forecasts for the host firm. The forecasts are published in a yearly or quarterly format complete with confidence boundaries for many companies. It is important to note that the market portfolio serves as a benchmark in the analysis and it need not summarize all the available investment alternatives.

If the host firm's sales forecasts are unsuitable or unavailable, an alternative is to use the forecasts of macroeconomic variables from a major econometric forecasting firm such as Data Resources Inc. or Wharton Econometrics. The forecasting consultants publish history and projections, with confidence boundaries, on various U.S. macroeconomic time series.

The covariance between the benchmark (the portfolio of investment alternatives) and the target investment can be calculated using a modification of the variance formula (Equation 8):

$$\text{Covariance of x,y} = \sigma_{x,y} = \frac{\sum_{i=1}^{N} (x_i - \mu_x)(y_i - \mu_y)}{N-1} \tag{9}$$

CAPM methodology benefits are in specifying the income stream and the confidence intervals. The CAPM forces accurate identification of the risks associated with each potential investment. In addition, the risk can vary for different investments. One potential investment may be as safe as the CD mentioned earlier, while another may be nothing more than a lottery ticket. The volatility differences between the two income streams are identified and incorporated into the model. There is no need to adjust the discount rate.

To compare different investments the CAPM model is used. The β's of the investments are calculated from the variance and covariance values:

$$\beta_i = \frac{\sigma^2_{i,Market}}{\sigma^2_{Market}} \tag{10}$$

The estimated β values calculated for the investments incorporate the risk, or uncertainty, of the investment through the variance measurements; and additionally, the timing of the investment stream is incorporated through the covariance measurements.

Investments with a calculated β greater than 1 have more risk than the benchmark portfolio, and those investments with a calculated β less than 1 are less risky. The magnitude of the calculated β reflects the volatility and timing of the return stream.

Although the uncertainty measurements may still involve a degree of subjectivity, the values for the risk must be directly specified in the analysis.

Two projects with differing probabilities of return were compared in Tables I and II of Section 1. Project 1 was favored over project 2 even though project 2 had a higher probability of return. It was assumed that the parameters of the projects were equal except for the probability of return, and the time-period where the project costs are incurred. The parameters of the CAPM analysis (R_f and μ_{Market}) are equivalent for the two projects, only the probable return ($\mu_{Project}$) differs. The β value is proportional to the expected return:

$$\beta_i = \frac{\mu_i - R_f}{\mu_{Market} - R_f} \tag{11}$$

A CAPM analysis of the two projects would favor project 2; $\beta_{Project\ 2} > \beta_{Project\ 1}$. This is because the expected return of Project 2 is greater than the expected return of Project 1. The other parameters used in the CAPM analysis (risk-free return R_f, and the average return in the market μ_{Market}) are identical between the projects.

4. AN EXAMPLE

Hercules Incorporated is composed of three separate companies: Specialty Chemicals, Aerospace, and Engineered Polymers. The profit rates of the different companies can be determined from the quarterly and annual reports to the shareholders. The annual reports are publicly available.

The goal of the example is to determine the structure of the Hercules portfolio, and to determine the contributions of the individual companies to the portfolio. The analysis is based on data from the published sources mentioned above, and the time period is the first quarter 1980 through the third quarter 1989. The data are not normalized into constant-year dollars because the calculations are based on profit rates. Table 4–1 presents a summary of the calculations.

Table 4–1 Profit Rate Statistics for Hercules Inc.

	HERCULES	SPECIALTY CHEMICALS	AEROSPACE	ENGINEERED POLYMERS
Mean	6.68%	8.78%	8.68%	5.06%
Maximum	9.99%	14.38%	14.58%	12.87%
Minimum	3.77%	1.40%	3.52%	-4.44%
Variance (x_i)	2.39%	8.31%	5.40%	17.06%
Covariance (x_i, Hercules)	1.00%	1.99%	1.09%	3.58%
Beta (x_i, Hercules)	1.00%	0.83%	0.46%	1.50%

The profit calculations are performed for the individual Hercules companies, and for the Hercules portfolio combining the three companies' profits. For the 1980:Q1 through 1989:Q3, the profit average, maximum, minimum, and variance are reported. The calculated β values are based on comparing the individual company to the Hercules portfolio. The Hercules portfolio β

is equal to 1 by definition. The β for the Specialty Chemical Company is calculated using Equation 6, (the covariance $_{i, \text{Hercules}}$ divided by variance$_{\text{Hercules}}$, or 1.99 / 2.39 = 0.83).

The focus of the analysis is to determine the contribution of the Aerospace company to the Hercules portfolio. The Hercules Aerospace profit rate, 5.40 percent, shows a greater dispersion than does the Hercules Corporate profit rate, 2.39 percent. The larger variance signifies that the volatility, or risk, for the Aerospace company is greater than for the Hercules portfolio of companies. Yet the calculated β value for the Aerospace company is less than one. The β value indicates that the profit rate of the Aerospace company is less cyclical than the Hercules portfolio of companies, where the profit rate of the Hercules Incorporated portfolio of companies defines the cycle. CAPM analysis cannot identify the source of the cycle. The CAPM technique highlights the cyclical characteristics for further research. Even though profits for Hercules Aerospace are volatile, the Aerospace profits contribute to the overall portfolio of companies by reducing the risk.

The contributions of the other companies can be calculated similarly, and the β calculations can be used to rank the investments.

The data are for the decade of the 1980s. During the 1980s there was a substantial increase in U.S. defense expenditures, and the increase contributed to the cyclical character of Hercules Aerospace. There are no reasons to expect the β values in Table III to remain constant during the decade of 1990s. The proper technique to extend the analysis into the future would be to use the company sales forecasts and confidence intervals, perhaps combining the sales forecasts with the projections of trends in U.S. industry.

If the traditional methods of analysis were followed (raising the discount rate to reflect the increased volatility of the Aerospace company profits), the traditional analysis would fail to capture the cyclical nature of the Aerospace Company. The traditional analytical techniques fail to incorporate all of the information available, and therefore may generate incomplete results.

5. CONCLUSION

The analytical framework of the Capital Asset Pricing Model can be adapted to capital investment analysis. The assumptions required for the analysis are not overly restrictive, nor are the computational demands. The most restrictive assumptions are the normal probability distribution, and independent errors in the per-period historical data or in the future estimates. The assumption of a normal distribution for the estimates can be substituted by the assumption of any stable Paretian distribution if the distributional parameters are known. The central limit theorem indicates that the assumption of a normal distribution will apply at the limit. The independence of errors in the data may also be a restrictive assumption, but if the correlation between the errors is known, then the correlation can also be incorporated into the model.

The advantages of the model are that alternative investments are compared on an equal basis; discount rate adjustments are not necessary. The CAPM avoids the bias towards myopic planning horizons inherent in the traditional methods that are used to incorporate uncertainty. The CAPM analysis also incorporates the covariance between the target income stream and the baseline income stream. The CAPM analysis can be combined with the traditional methods of analysis to provide a more complete assessment of project value.

The theory of CAPM analysis is taught in many business school programs, and many capital analysts are familiar with the technique as applied to stocks and bonds. CAPM analysis can be extended to capital projects without difficulty.

REFERENCES

1. Fama, E. F., *Foundations of Finance*, New York: Basic Books, 1976.

2. Garbade, K. D., *Securities Markets*, New York: McGraw–Hill, 1982.

3. Larson, H. J., *Introduction to Probability and Statistical Inference*, Third Edition, New York: John Wiley, 1982.

4. Wilson, R. "Risk Measurements in Public Projects," in R. Lind (editor), *Discounting for Time and Risk in Energy Policy*, Resources for the Future Inc. 1982, 205–209.

A METHODOLOGY FOR ASSESSING SCHEDULE RISK IN SOFTWARE DEVELOPMENT ESTIMATES

by

William M. Stein
The MITRE Corporation
Bedford, MA 01730

ABSTRACT

Planning estimates for software development schedules are often obtained using a schedule-estimating equation that expresses the schedule estimate (in calendar months) as a nonlinear function of the total development effort (in staff-months). For example, the COCOMO schedule model [1] for embedded-mode developments is

$$\text{Schedule} = 2.5\ (\text{Effort})^{0.32}$$

We address the problem of measuring the error of uncertainty in the schedule estimate when the error of the effort estimate and the error of the schedule model (given actual effort data) are known.

We illustrate the methodology for two models calibrated to historical cost data.

1. INTRODUCTION

The General Problem

A widely accepted method for estimating the time necessary for a software development is to start with an effort estimate (in staff-months) and compute the schedule, i.e., the length of time it will take to complete the project (in months), via an estimating relationship of the form

$$S = aE^b$$

where

E	=	software development effort in staff–months
S	=	the corresponding schedule, in calendar months

and the exponent b generally ranges from 0.25 to 0.38. [1] Hence, such schedule estimating relationships are informally referred to as "cube root formulas."

B. W. Boehm's COCOMO model [1] uses the following relationship to estimate schedule for embedded–mode[1] developments:

$$S = 2.5\ E^{0.32}$$

Recalibrating this formula specifically for embedded software projects of the U.S. Air Force's Electronic Systems Division (ESD), the MITRE Cost Analysis Technical Center has obtained the relationship [2]

$$S = 3.7\ E^{0.32}$$

For non–embedded software developments in the Ada[2] programming language, the MITRE Cost Analysis Technical Center has statistically fit this formula [3]:

$$S = 5.9\ E^{0.23}$$

Schedule estimation using these models first requires an estimate of software development or modification effort. An effort estimate is typically generated from a parametric model, with considerable uncertainty in the inputs, e.g., number of delivered source instructions and cost driver attributes. What a cost analyst or project manager desires of a model is a measure of uncertainty or "risk" associated with a given estimate, from which "confidences" can be stated numerically. A practical uncertainty analysis proceeds from a probability distribution of possible outcomes (schedules, in this case). Using such a distribution, a project manager can derive the

---

[1]The "embedded" mode of software development characterizes a highly constrained development environment. A general definition for embedded–mode software development is a development to pre–existing constraints which are usually tight, strongly coupled to the external hardware system and to operational procedures, with considerable need for innovation, and a high premium on early completion.

[2]Ada is a registered trademark of the U.S. Government (Ada Joint Program Office).

probability that an estimated schedule is not exceeded, versus any hypothetical "actual" schedule. See Fig. 1–1 for an example.

The general problem addressed here is thus to combine a description of input effort uncertainty with a description of schedule model uncertainty, and provide a probability distribution of actual schedule that may result.

We provide an easy-to-use table look-up procedure for obtaining the probability that an estimated schedule is not exceeded. The methodology employed was suggested by K. W. Pullen [4], who carried out a more abbreviated uncertainty analysis on Boehm's original COCOMO schedule model, the first of the equations presented above.

Effort Uncertainty

From a single number for the effort to be required of a software development or modification, be it a methodical calculation or a rough order-of-magnitude "educated guess," there are essentially two different ways to frame one's uncertainty as a probability distribution about that point estimate. Early in the planning phases of a project, a gross characterization of uncertainty is appropriate, whereas once requirements are firmer, the more specific causes of the uncertainty tend to become recognizable.

In the early phases of a project's development, such as during feasibility studies, requirements analyses, or preliminary design activities, uncertainty tends to range over multiples (or fractions) of the "most likely" estimate. To illustrate this, Fig. 1–2 is taken from [1].[3] We will assume that the project planner believes the effort estimate less likely to land near the extremes than near the current "best guess." Thus an appropriate parameterized family of distributions about an effort estimate E is the triangular distribution,[4] from a minimum of E/c to a maximum at cE, with mode (i.e., the peak of the distribution) at the given estimate E, where c represents

[3]Barry W. Boehm, *Software Engineering Economics*, ©1981, p. 311. Reprinted by permission of Prentice-Hall, Inc., Englewood Cliffs, NJ.

[4]Both the triangular and beta distributions are often used in risk analyses [5, 6, 7]. The triangular distribution captures skewness and decreasing probability in the tails sufficiently for our purposes, and permits organizing our tabulated results via single parameter look-up, whereas the beta is a two-parameter distribution.

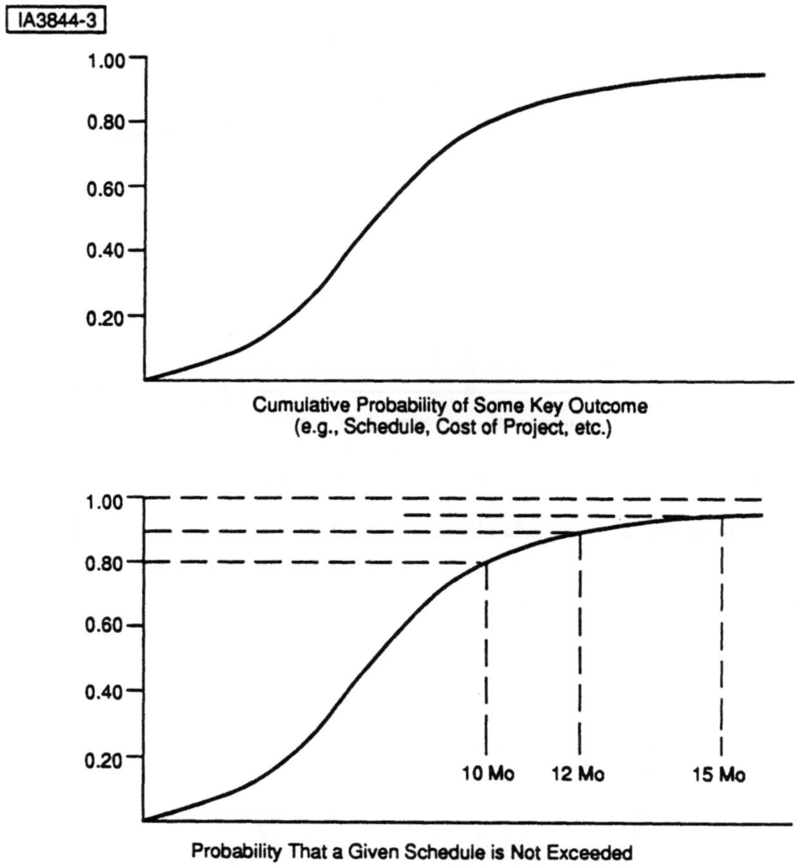

Confidence	Risk	Schedule
80%	20%	10 Months
90%	10%	12 Months
95%	5%	15 Months

Figure 1-1 Example of a Schedule Estimate Uncertainty Analysis Utilizing a Cumulative Probability Distribution

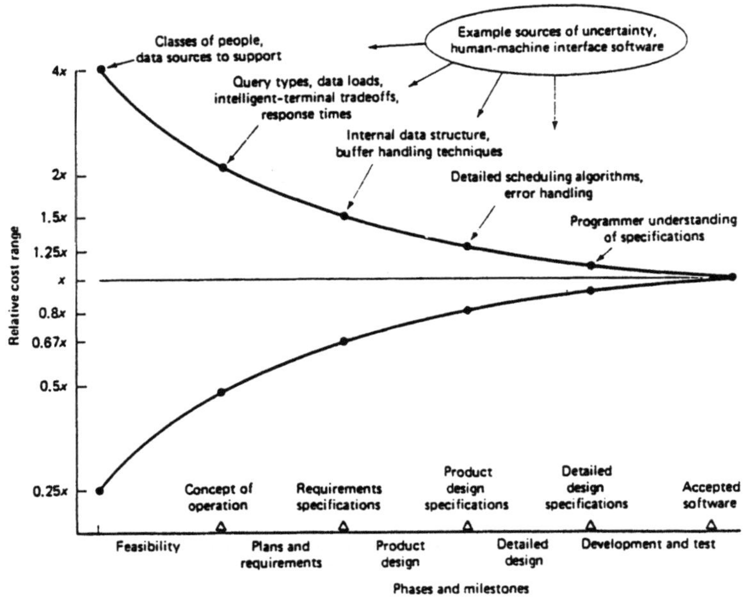

Figure 1–2 Software Cost Estimation Accuracy versus Phase [1]

an uncertainty multiplier. For such a family of triangular distributions, values of the uncertainty multiplier used in this work are c = 1.5, 2.0, 3.0, and 4.0. The resulting family of effort distributions is illustrated in Fig. 1–3.

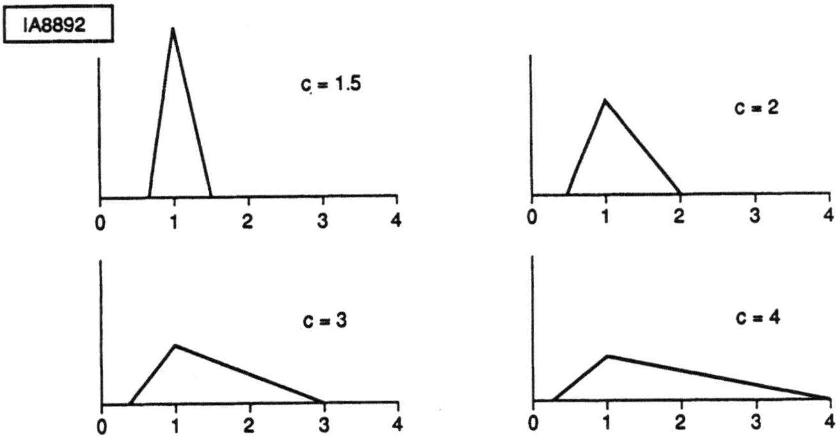

Figure 1–3 Triangular Distributions Used for Effort Input

When a software project is understood sufficiently well for a number of its specific components to be recognized as sources of uncertainty about their predicted effort estimates, and when the overall effort is a sum of these component efforts, then the normal distribution is a justifiable choice [7]. In contrast to the triangular family of distributions discussed above, deviations are symmetric for equal values above and below the central point estimate. To permit standardization about any mean, the coefficient of dispersion D, defined as the ratio of standard deviation (sigma) to mean, will be taken as the relevant parameter. Values used in this work are D = 1/12, 1/6, 1/4, and 1/3. These normal distributions are illustrated in Fig. 1–4.

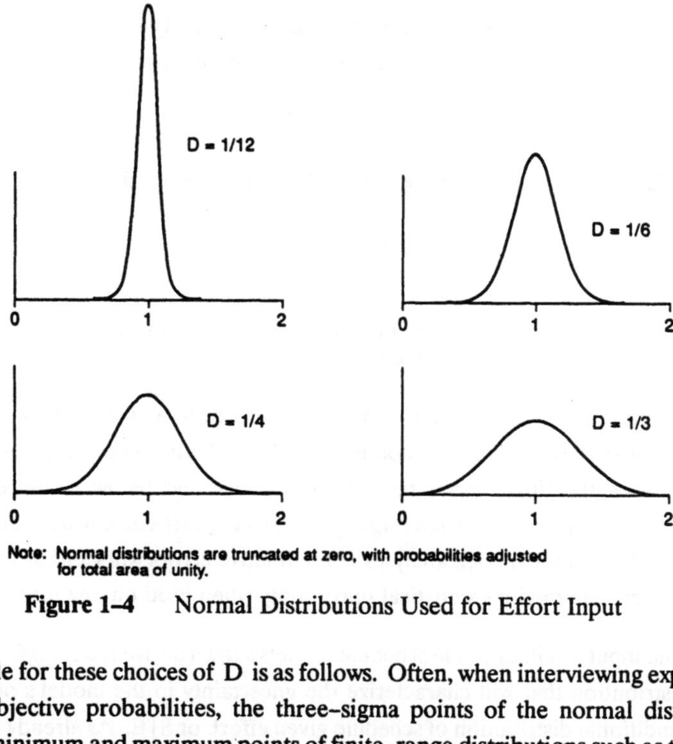

Note: Normal distributions are truncated at zero, with probabilities adjusted for total area of unity.

Figure 1–4 Normal Distributions Used for Effort Input

Rationale for these choices of D is as follows. Often, when interviewing experts in order to establish subjective probabilities, the three–sigma points of the normal distribution are likened to the minimum and maximum points of finite–range distributions such as the triangular or beta. Asking a software analyst to judge whether a practical minimum bound to an effort estimate extends one quarter of the way down to zero, half way, three–quarters of the way or all the way to zero is equivalent to choosing among D = 1/12, D = 1/6, D = 1/4, or D = 1/3, respectively.

The tails of the normal distribution extend from negative infinity to infinity. Since effort cannot be less than zero, we truncate the effort distribution for such values, and adjust the height of the distribution accordingly, to preserve total area of unity. For the case $D = 1/3$, this truncation occurs precisely three standard deviations below the mean, E. (The positive tail, from E plus three sigma to infinity, need not be truncated.)

2. SCHEDULE MODEL UNCERTAINTY

Two schedule models are utilized in this uncertainty analysis -- the ESD/MITRE recalibrated COCOMO equation for embedded-mode development schedules,

$$S = 3.7\, E^{0.32}$$

and the MITRE equation for non-embedded Ada development schedules,

$$S = 5.9\, E^{0.23}$$

where

E = software development or modification effort in staff-months
S = the corresponding schedule, in months.

Each of these formulas is obtained by a statistical fit to a given database of projects for which actual software development or modification schedules are known. Since these points do not lie precisely on the fitted curve, but fall both above and below it, there is a statistical uncertainty, or modeling error, surrounding any one project's schedule when estimated by such a formula. For each schedule model, analyses of the relative errors will yield probability distributions that represent the model's statistical uncertainty when used on any future project.

Since the input to either of the schedule models shown above is a *specific* value for E, the probability distribution that will characterize the uncertainty in the model's output S must in reality be a conditional distribution of schedule given effort, or $S \mid E$. As already noted, development of a distribution for $S \mid E$ begins with a study of the relative errors produced by the schedule model.

For our purposes, we define[5]:

$$\text{Relative Error} = \frac{\text{Actual Schedule} - \text{Estimated Schedule}}{\text{Estimated Schedule}}$$

For the ESD/MITRE recalibrated COCOMO equation for embedded software, Fig. 2–1 shows the actual schedule vs. effort data points used for the recalibration (which include modification projects as well as new software developments) with the recalibrated estimating equation (model) superimposed on this actual data. Figure 2–2 displays only the relative error (as we have defined it) against the same horizontal axis as that of Fig. 2–1.

IA8703-2

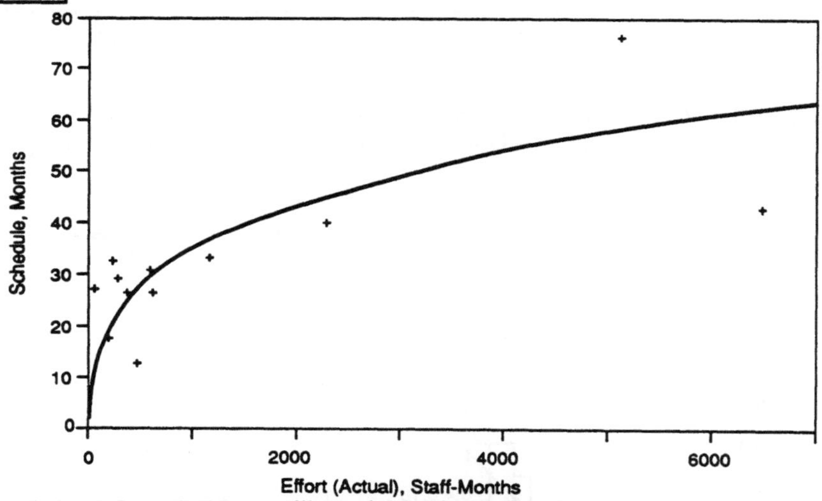

Figure 2–1 ESD/MITRE Recalibrated COCOMO Schedule, Embedded Mode Projects

The step function in Fig. 2–3 is the empirical cumulative distribution function for the relative error data of Fig. 2–2, where these data have been interpreted as twelve independent observations of a single random variable. Superimposed over this empirical distribution is a specific cumulative distribution of the beta family (conforming to the requirement that it be finite

[5]This is an inversion of the usual definition of relative error, (estimated – actual)/actual. In our situation, we "know" the estimate; the actual is the unknown quantity, and so we define relative error with respect to the estimate.

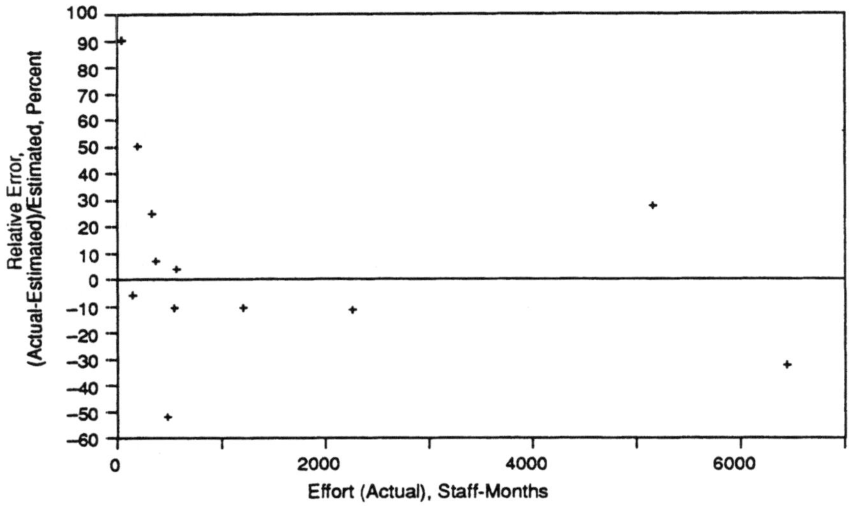

Figure 2–2 Relative Error of Schedule Estimate vs. Effort for ESD/MITRE Recalibrated COCOMO Schedule Model, Embedded Mode Projects

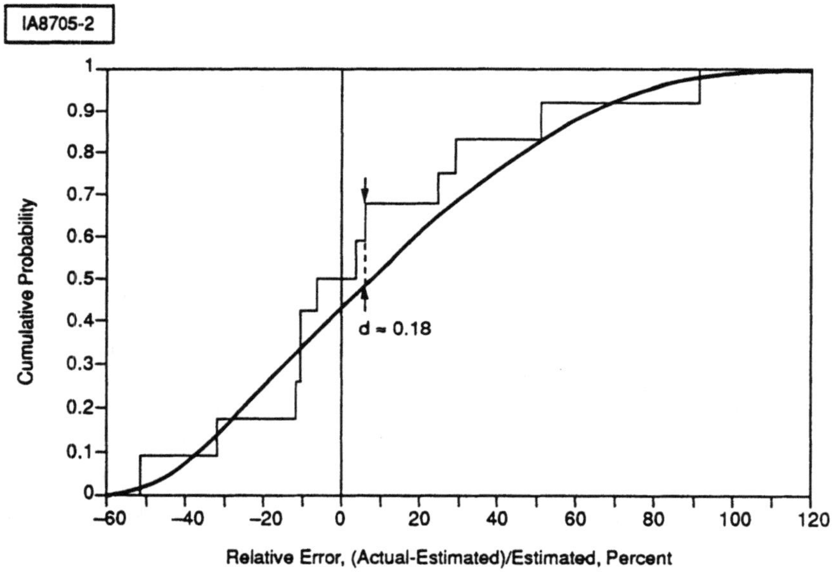

Figure 2–3 Beta Distribution Fit to Empirical Cumulative Distribution Function of Relative Error in ESD/MITRE Recalibrated COCOMO Schedule Model

to the left, since no schedule can be less than zero, and hence no relative error can be less than −100 percent). The maximum displacement between the two, d ≈ 0.18 as noted in the figure, shows this particular beta distribution to be an acceptable fit in the sense that it is not rejected by the Kolmogorov–Smirnov goodness-of-fit test. The beta density function of S|E for this recalibrated COCOMO model is that illustrated in Fig. 2–4.

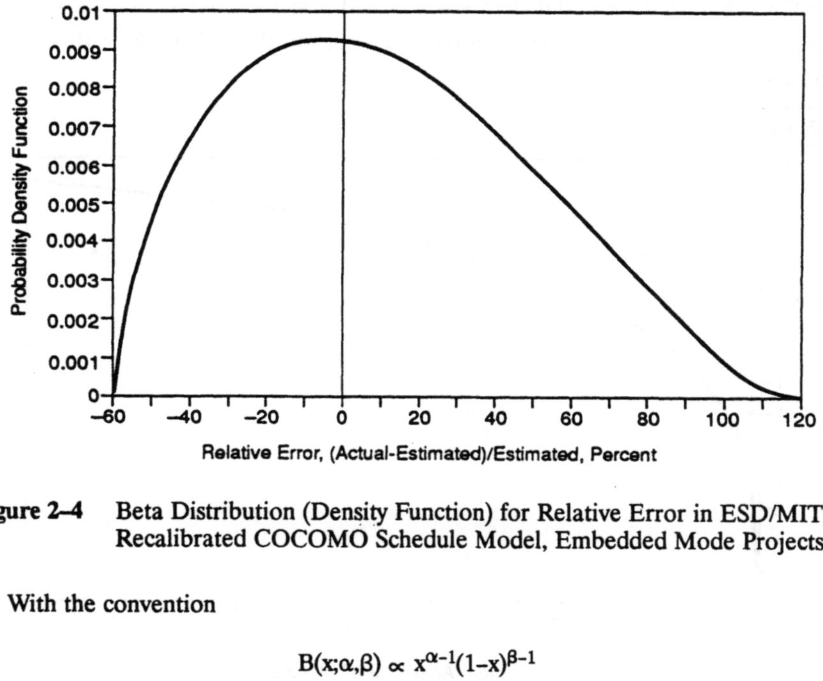

Figure 2–4 Beta Distribution (Density Function) for Relative Error in ESD/MITRE Recalibrated COCOMO Schedule Model, Embedded Mode Projects

With the convention

$$B(x;\alpha,\beta) \propto x^{\alpha-1}(1-x)^{\beta-1}$$

the shape parameters of the selected beta distribution are $\alpha = 1.72$ and $\beta = 2.66$. The standardized range of (0, 1) is then rescaled to (−60 percent, 120 percent) for relative error, which implies a rescaling to (0.4, 2.2) for the ratio of actual schedule to estimated schedule.

We now have determined the *relative* error distribution, incorporating data from a variety of projects of different sizes and schedules; it translates to a distribution of *actual* schedule by addition of 1.0 and multiplication by the estimated schedule.

For the model applicable to non-embedded software developments in Ada, the corresponding actual and estimated data, and relative errors, are shown in Figs. 2–5 and 2–6. The step function in Fig. 2–7 is the empirical cumulative distribution function for the relative error data of Fig. 2–6, where these data have been interpreted as nine observations of a single random variable. Figure 2–7 also displays a specific beta distribution fit to this data, and its associated Kolmogorov–Smirnov test statistic, $d \approx 0.15$. Finally, the beta distribution of $S \mid E$ for this model is given in Fig. 2–8, with shape parameters $\alpha = 1.22$ and $\beta = 1.48$, and the standard range of (0, 1) rescaled to (–20 percent, 30 percent) in relative error, or (0.8, 1.3) in the ratio of actual to estimated schedule.[6]

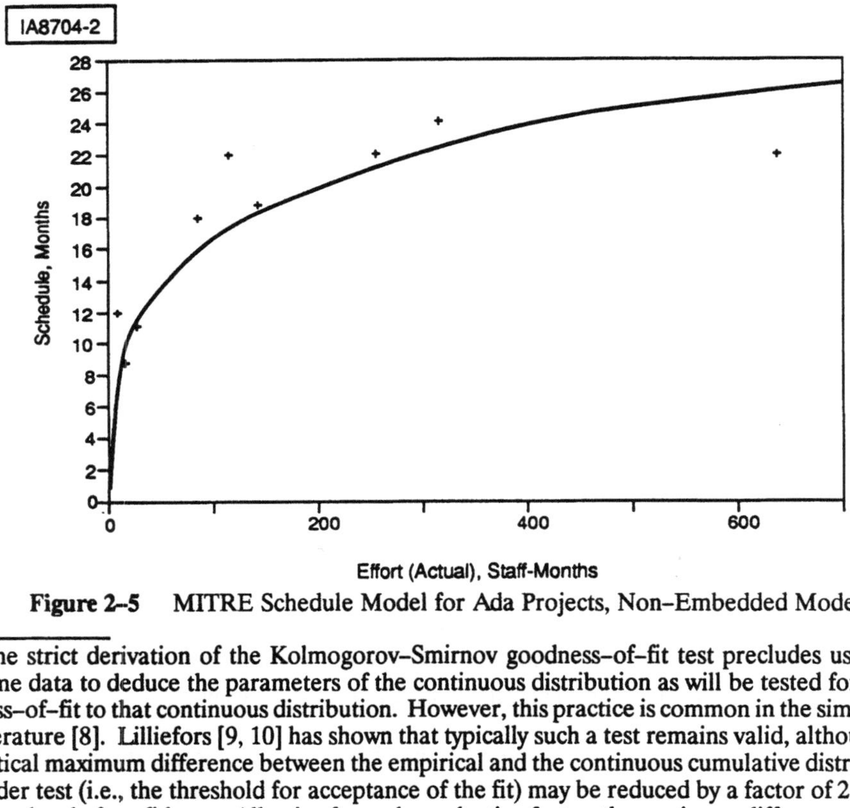

Figure 2–5 MITRE Schedule Model for Ada Projects, Non-Embedded Modes

[6]The strict derivation of the Kolmogorov–Smirnov goodness-of-fit test precludes using the same data to deduce the parameters of the continuous distribution as will be tested for goodness-of-fit to that continuous distribution. However, this practice is common in the simulation literature [8]. Lilliefors [9, 10] has shown that typically such a test remains valid, although the critical maximum difference between the empirical and the continuous cumulative distribution under test (i.e., the threshold for acceptance of the fit) may be reduced by a factor of 2/3 for a given level of confidence. Allowing for such a reduction factor, the maximum differences shown in Figs. 7 and 11 will still support that their respective beta distributions cannot be rejected even at the 10 percent level of significance.

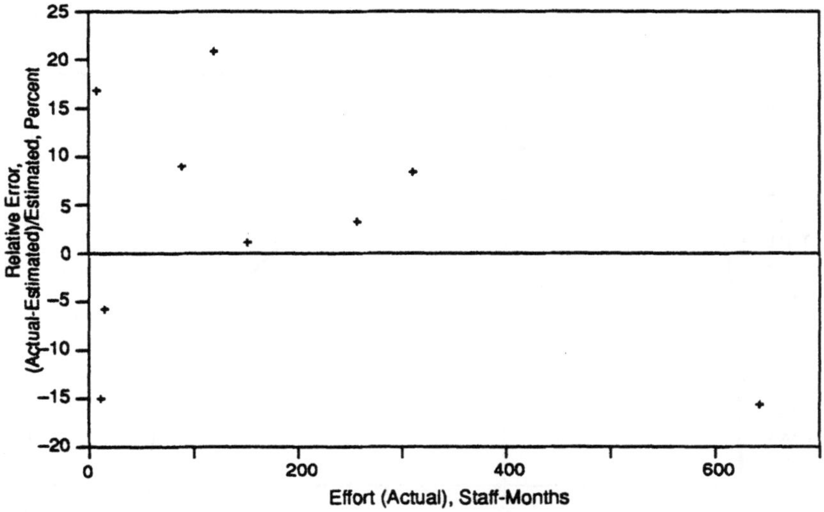

Figure 2–6 Relative Error of Schedule Estimate vs. Effort for MITRE Schedule Model for Ada Projects, Non–Embedded Modes

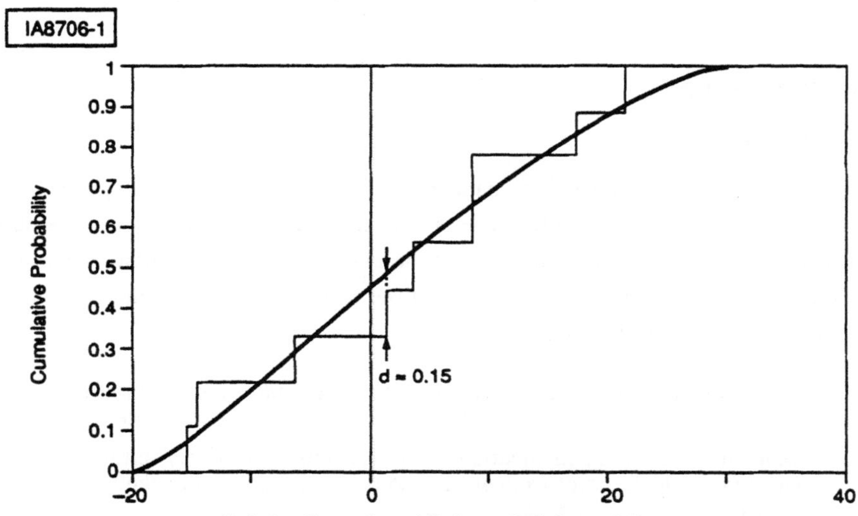

Figure 2–7 Beta Distribution Fit to Empirical Cumulative Distribution Function of Relative Error in MITRE Schedule Model for Ada Projects

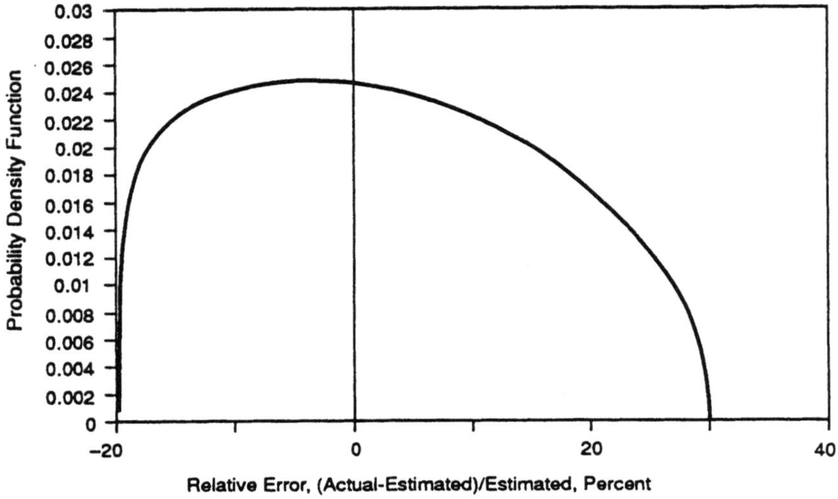

Figure 2–8 Beta Distribution (Density Function) for Relative Error in MITRE Schedule Model for Ada Projects, Non–Embedded Modes

3. METHODOLOGY

Monte Carlo Simulation

To reach the objective of this work, namely an easy–to–use table look–up procedure for risk or probability that an estimated schedule is not exceeded, it is necessary to develop the probability distribution for schedule that should be placed about the resultant single point schedule estimate S_0 one obtains when using the given schedule model and the "most likely" effort estimate E_0. Inputs to the process are:

(1) a description of software development effort uncertainty (in staff–months) in the form of a probability distribution about a "most likely" estimate

(2) a schedule model (i.e., an equation) to convert this "most likely" effort estimate into a single point estimate for software development schedule (in months)

(3) a probability distribution capturing the uncertainty among the data from which the schedule model was derived.

These inputs were described in the previous section.

Formally, the problem is as follows. Given a probability density function $p_E(x)$ for effort, a schedule model equation $S = aE^b$, and a conditional probability density function $p_{S|E}(y|x)$, find the unconditional probability density function

$$p_S(y) = \int p_{S|E}(y|x)p_E(x)dx$$

The density function $p_E(x)$ will have mode at E_o and values of nonzero probability ranging from some $x = E_{min}$ to some $x = E_{max}$ (including the range from zero to infinity, as in the case of the truncated normal distribution). Values of nonzero probability for $p_S(y)$ will range from $y = S_{min}$ to $y = S_{max}$ (including the range zero to infinity), where S_{min} is the start of the nonzero probability for $p_{S|E}(y|E_{min})$, and S_{max} is the upper end of the interval of nonzero probability for $p_{S|E}(y|E_{max})$.

The above integral cannot be worked out in general, and can be solved numerically for certain cases only with difficulty. One case that should appear mathematically simple, for example, is the case where both $p_{S|E}(y|x)$ and $p_E(x)$ are triangular. This dual-triangular case is comprised of separate definitions of the integrand for eight different subintervals on the y-axis. All but the first and last subinterval require piecewise integration (along the x-axis) of two or three components. This integration is between boundaries, some of which follow curves proportional to $y = ax^b$. As specific numerical inputs vary for different choices of triangular probability density functions, the sequence of integrations may vary due to the manner in which the boundary intersections eclipse each other.

The case where these density functions are normal (and not truncated at zero), easy to work with in closed form in other settings, yields intractable integrals for this problem. The difficulty is that whereas the mean of $p_E(x)$ is the mean effort , E_o, the mean of $p_{S|E}(y|x)$ is the result of the schedule model for the given x, i.e., ax^b. Thus one does not have the familiar quadratic forms in the exponents of the integrand.

To obtain an unconditional probability density function of schedule $p_S(y)$ "centered" in an appropriate way on $S_o = aE_o{}^b$, a two-stage Monte Carlo computer simulation was developed and carried out. First, depending upon the desired choice of family of effort distributions and shape or endpoint parameter(s), a random number E was drawn from a probability distribution representing effort. Next, a second random number S was drawn from the appropriate beta distribution representing schedule modeling error for the desired choice of schedule model. For the ESD/MITRE recalibrated COCOMO schedule model for embedded software projects, the beta distribution had shape parameters $\alpha = 1.72$ and $\beta = 2.66$, with endpoints

$$S_{min} = 0.4\ (3.7\ E^{0.32}), \qquad S_{max} = 2.2\ (3.7\ E^{0.32})$$

For non-embedded Ada software developments, the shape parameters were $\alpha = 1.22$ and $\beta = 1.48$, and the endpoints of the beta distribution were

$$S_{min} = 0.8\ (5.9\ E^{0.23}), \qquad S_{max} = 1.3\ (5.9\ E^{0.23})$$

A fine-grained histogram was built up from these S values, from which percentiles and other statistics were found.

For each histogram, the number of E and S random variables computed was 500,000 each, for a total of 1,000,000 calls to a random number generator. At this sample size, the sensitive 1st, 5th, 95th and 99th percentiles were negligibly affected by a re-running of the program with a different starting seed. In all cases, an underlying uniform random number generator was actually used, and conversion to the desired probability distribution was accomplished by the inverse probability integral transformation

$$Y = F_Y^{-1}(U)$$

where

U = represents a uniformly distributed random variable
Y = the random variable with desired cumulative distribution function $F_Y(y)$.

4. STANDARDIZATION OF SCHEDULE DISTRIBUTIONS

The set of schedule distributions that results from this analysis and Monte Carlo simulation is generic to any planning schedule estimate regardless of the level of software development or modification effort involved. Effort uncertainty need only conform to the shape/width specifications of the families of input distributions selected for this work; the output distributions are standardized (normalized) so as to be applicable to any single point estimate of schedule. For each effort distribution input to the Monte Carlo computer program, the "most likely" effort (i.e., the mode of the distribution) was put at that number E_o such that $S_o = 100$, i.e.,

$$E_o = (100/3.7)^{1/0.32}$$

for the ESD/MITRE recalibrated COCOMO schedule model for embedded software projects, and

$$E_o = (100/5.9)^{1/0.23}$$

for non–embedded Ada software developments.

The resulting $S_o = 100$ is to be interpreted not as 100 months, but as 100 *percent* of whatever schedule is calculated to be a point estimate. Thus all points below and above $S_o = 100$ are to be interpreted as percentages of the estimated project schedule that carry the corresponding probability of not being exceeded (see Fig. 1).

Such standardization (normalization) can be put to this Monte Carlo analysis because, just as only the shape of the selected effort distribution need be considered as input, so only the shape of the computed schedule distribution need be interpreted as output. Formal justification derives from the fact that the function $y = ax^b$ is monotonic and invertible. Details are given in an appendix.

From the discussion in the appendix, it will be noted that S is in reality a *product* of two independent random variables, and the two–stage Monte Carlo process could have been implemented accordingly:

(1) Draw two random numbers

(2) Convert one to a random variable T whose distribution has the same shape as that of E but rescaled for a mode at 1, and

(3) Convert the other to B', a beta variable having the appropriate shape parameters and rescaled to represent relative error.

(4) Then form the product $S = 100\ T^{0.32}B'$.

The implementation as discussed earlier was selected, however, for intuitive value and readability of the FORTRAN program.

We note that there is potentially an analytic solution to this problem. Since the desired schedule variable S is a product of two random variables, one of which is an underlying effort variable raised to a power, the Monte Carlo solution approach could be replaced by a completely analytic process when the effort distribution is assumed to have a lognormal distribution. It would be necessary to fit an independent lognormal distribution (in place of the beta distribution) to the relative errors. The distribution of S, as a product of independent lognormal distributions, would then also have a lognormal distribution.

5. RESULTS

Application of Tabulated Results

The results of the Monte Carlo simulation for two schedule models, two input families of effort distributions per schedule model, and four specific cases per input family of distributions, are presented in Tables 5–1 through 5–4. The table entries are in percentages, where the point estimate produced by the given schedule model $S = aE^b$ corresponds to 100 percent. The schedule distribution for each case is given as a set of percentiles, the Pth percentile being defined as that percentage of the point estimate schedule for which the probability of project completion in as much or less time is $p = P/100$. The probability that the project will require more time than that of the Pth percentile is thus 1–p.

Each table also provides, for each case, those probabilities $p = P/100$ for which the Pth percentiles take the values of 99 percent, 100 percent, and 101 percent of the schedule estimate. Since 100 percent corresponds to the underlying point estimate of schedule, the probability at S = 100 percent is actually the probability of the software development or modification taking not more than the estimated amount of time. Probabilities at 99 percent and 101 percent (of the schedule estimate) are shown, to provide a measure of sensitivity about the 100 percent point.

These tables may be used to answer a variety of questions, as illustrated by the following example. Suppose a software development is budgeted based upon an estimated 500 staff-months. The software will be developed in the embedded mode, in one of the older DOD–approved programming languages (not in Ada). Thus, the development schedule is estimated by the ESD/MITRE recalibrated COCOMO equation as $3.7(500)^{0.32} = 27.0$ months. What is the probability that this schedule will be exceeded? For what schedule can a project manager be 90

Table 5–1 Software Development Schedule Percentiles for ESD/MITRE Embedded Projects, Effort Distribution Triangular

Schedule Model: $S = 3.7\ E^{0.32}$

Schedule Corresponding to Most Likely Effort: 100%

Uncertainty Multiplier	1.5	2	3	4
Theoretical Min.	35.13%	32.04%	28.14%	25.67%
Percentile				
0	35%	32%	29%	26%
1	44%	45%	45%	46%
5	54%	55%	57%	59%
10	62%	63%	66%	69%
20	76%	77%	81%	84%
30	87%	89%	94%	98%
40	99%	101%	106%	111%
50	110%	112%	118%	124%
60	121%	124%	131%	138%
70	134%	137%	145%	153%
80	148%	153%	162%	172%
90	167%	173%	186%	198%
95	182%	189%	205%	220%
99	204%	215%	238%	258%
100	244%	266%	302%	331%
Prob. @ S = 99%	0.404	0.385	0.344	0.309
Prob. @ S = 100%	0.413	0.394	0.352	0.317
Prob. @ S = 101%	0.422	0.402	0.361	0.325
Mean Schedule	113%	116%	123%	130%
Standard Deviation	39%	41%	46%	50%
Theoretical Max.	250.48%	274.63%	312.68%	342.83%

percent confident that there will not be an overrun in elapsed time (equivalently, the schedule that has only a 10 percent risk of not being met)?

To answer these questions, let us first assume that the original 500–staff–month estimate might actually be as little as 250 staff–months or as much as 1,000 staff–months, as suggested by Fig. 1–2 for the early planning and requirements phases. Thus Table 5–1 is selected, with an uncertainty multiplier of 2. The probability that the estimated 27 months will be exceeded is 1 minus 0.394 (this being the probability at S = 100 percent for the second column of the table), or 0.61. To be 90 percent confident that a schedule will be met (i.e., for a 10 percent level of risk), the project manager must plan for 173 percent of 27 months, or 47 months.

Table 5–2 Software Development Schedule Percentiles for ESD/MITRE Embedded Projects, Effort Distribution Normal

Schedule Model: $S = 3.7\ E^{0.32}$

Schedule Corresponding to Most Likely Effort: 100%

Coeff. of Disperson	1/12	1/6	1/4	1/3
Percentile				
1	44%	44%	43%	40%
5	53%	53%	52%	51%
10	61%	61%	60%	59%
20	75%	74%	73%	72%
30	86%	86%	85%	84%
40	97%	97%	96%	95%
50	108%	108%	107%	106%
60	119%	119%	118%	117%
70	132%	131%	131%	130%
80	146%	146%	145%	145%
90	165%	165%	165%	165%
95	178%	179%	180%	181%
99	199%	201%	204%	208%
Prob. @ S = 99%	0.416	0.420	0.428	0.439
Prob. @ S = 100%	0.425	0.430	0.437	0.448
Prob. @ S = 101%	0.435	0.439	0.446	0.457
Mean Schedule	111%	111%	111%	110%
Standard Deviation	38%	39%	39%	40%

Further questions concerning this example, which can be answered by simple table look–up procedures, are the following. Can one find lower and upper bound schedules defining a 90 percent confidence interval about the 27–month estimate? Could the 27–month number be referred to as a mean?

Again, we find our answers in Table 5–1. For a 90 percent confidence interval about the original point estimate of 27 months, select the 5th and 95th percentiles: 55 percent and 189 percent respectively, which translate to lower and upper bound schedules of 15 and 51 months, respectively. The 27–month point estimate is somewhat below the mean, which Table 5–1 puts at 116 percent, or 31 months.

As with any statistical justification, the more "confidence" one demands, the wider becomes the interval that assures it. 100 percent confidence is achieved only by taking on the entire range of possible outcomes. If the project manager for the above example were willing to

Table 5-3 Software Development Schedule Percentiles for Ada Non-Embedded Projects, Effort Distribution Triangular

Schedule Model:	$S = 5.9\ E^{0.23}$			
Schedule Corresponding to Most Likely Effort:	100%			
Uncertainty Multiplier	1.5	2	3	4
Theoretical Min.	72.88%	68.21%	62.14%	58.16%
Percentile				
0	72%	68%	62%	58%
1	78%	77%	75%	75%
5	82%	83%	83%	84%
10	85%	86%	88%	90%
20	90%	92%	94%	97%
30	95%	96%	100%	103%
40	99%	101%	104%	108%
50	103%	105%	109%	113%
60	107%	109%	114%	119%
70	112%	114%	119%	125%
80	117%	119%	126%	132%
90	123%	126%	134%	141%
95	127%	131%	141%	149%
99	133%	140%	151%	161%
100	142%	152%	167%	178%
Prob. @ S = 99%	0.396	0.360	0.287	0.232
Prob. @ S = 100%	0.420	0.383	0.308	0.249
Prob. @ S = 101%	0.444	0.406	0.328	0.266
Mean Schedule	104%	106%	111%	115%
Standard Deviation	14%	15%	18%	20%
Theoretical Max.	142.71%	152.47%	167.37%	178.82%

live with a 30 percent level of risk instead of a 10 percent risk, then 173 percent could be replaced by 137 percent, and a 37-month schedule could be proposed instead.

6. OBSERVATIONS AND CONCLUSIONS

The 50th percentile is also known as the median of a distribution. In all cases in Tables 5-1 through 5-4, the median is greater than 100 percent (of the estimated schedule), and the mean is greater than the median, due to these schedule distributions being skewed to the right. The source of this skewness may be the effort distribution itself, such as those of Tables 5-1 and 5-3.

Table 5–4 Software Development Schedule Percentiles for Ada Non–Embedded Projects, Effort Distribution Normal

Schedule Model: $S = 5.9\ E^{0.23}$

Schedule Corresponding to Most Likely Effort: 100%

Coeff. of Disperson	1/12	1/6	1/4	1/3
Percentile				
1	79%	77%	73%	67%
5	82%	81%	80%	77%
10	85%	84%	83%	82%
20	90%	89%	89%	88%
30	94%	94%	93%	92%
40	98%	98%	97%	97%
50	102%	102%	102%	101%
60	106%	106%	106%	105%
70	111%	110%	110%	110%
80	115%	115%	115%	116%
90	121%	121%	122%	123%
95	125%	125%	126%	128%
99	129%	131%	134%	137%
Prob. @ S = 99%	0.419	0.427	0.440	0.454
Prob. @ S = 100%	0.443	0.451	0.464	0.476
Prob. @ S = 101%	0.468	0.476	0.487	0.500
Mean Schedule	103%	103%	103%	102%
Standard Deviation	13%	14%	14%	16%

However, in the case of the schedule model for ESD/MITRE embedded projects, right skewness is also due to the distribution of model error (Fig. 2–4).

Where the input distributions have finite range, Tables 5–1 and 5–3, the theoretical schedule minima and maxima are given. In these tables, the 0th and 100th percentiles that result from the Monte Carlo simulation are provided as a display of the accuracy of the simulation in the tails of the distributions. Generally, the longer but heavier tails require a greater number of random observations to establish the more sensitive 1st, 5th, 95th and 99th percentiles to a given degree of accuracy. The worst case for a right–hand tail is in the schedule resulting from a triangular distribution of effort, with the ESD/MITRE embedded software schedule model (Table 5–1) and an uncertainty multiplier of 4. The heaviest tails are possibly the left–hand tails of schedules whose input distributions are normal with coefficient of dispersion 1/3 (and thus noticeably truncated at zero), Tables 5–2 and 5–4.

In Tables 5–2 and 5–4, the cases of normally distributed effort with differing coefficients of dispersion produced schedule distributions that were closer to one another than occurred in the remaining tables. The reason for this is that, for the cases of D = 1/12, 1/6 and 1/4, the input distribution is narrower than the distribution of schedule modeling error (the beta distributions of Figs. 2–4 and 2–8), hence the output schedule distributions reflect schedule modeling error more than input effort uncertainty.

The triangular distribution with c = 1.5 and the normal distribution with D = 1/6 have nearly the same means (1.056 and 1.000 respectively) and standard deviations (0.171 and 0.167 respectively). A comparison of the first column of Table 5–1 with the second column of Table 5–2 (also, of the first column of Table 5–3 with the second column of Table 5–4) shows that the schedule distributions which result from these two different effort distributions are very nearly the same. As noted in the preceding paragraph, these outputs are similar not so much from inherent robustness in the schedule distribution estimation procedure but because the input effort distributions are narrower than the (beta) distributions used for schedule modeling error. The third and fourth columns of the tables illustrate how such robustness declines as one moves to wider effort distributions, for which the triangular distributions become increasingly skewed while the truncated normal distributions begin to exhibit only a minute degree of skewness (due solely to the truncation).[7]

To assess and control the accuracy of these results, each column of each table was created by averaging the results of two independent simulation runs of the Monte Carlo program. Each run produced 500,000 random observations per schedule distribution (by calling on the random number generator 1,000,000 times each, due to the two–stage nature of selecting a random effort before computing a random schedule). These runs differed only in their independent starting seeds. The random number generator was tested for uniformity (chi–square test and a limited amount of runs testing) and found to be free of cycles for at least 2,000,000 calls for each of the starting seeds used. Based upon analysis of the raw data of these simulation runs, and preliminary runs with 50,000 observations per schedule distribution, we feel that the numbers given in Tables 5–1 through 5–4 are accurate to $\pm$ 1 percent.

[7]Since the triangular distribution is often used in risk analyses as a simplified substitute for the beta distribution [5, 6, 7], it would be of interest to experimentally ascertain the robustness of our methodology to the choice between these two distributions at comparable means, standard deviations, and coefficients of skewness.

APPENDIX: FORMAL JUSTIFICATION FOR MULTIPLICATIVE SCALING

Multiplicative scaling justifies producing, via Monte Carlo simulation, effort and schedule distributions about any desired model value. In this work, the value of 100 has been selected to permit interpretation as a percentage. Such standardization (normalization) can be put to this Monte Carlo analysis because, just as only the shape of the selected effort distribution need be considered as input, so only the shape of the computed schedule distribution need be interpreted as output. Formal justification derives from the fact that the function $y = ax^b$ is monotonic and invertible.

Proof: (For ease of illustration, the specific values dictated by the ESD/MITRE recalibrated COCOMO equation will be used instead of generic variables.) Given that effort E (a random variable) takes the value x,

$$\Pr(S \leq s \mid E = x) = \Pr(B_x \leq s)$$

where B_x has a beta distribution with shape parameters $\alpha = 1.72$ and $\beta = 2.66$, with nonzero probability over the interval $(0.4 * 3.7x^{0.32}, 2.2 * 3.7x^{0.32})$. By rescaling,

$$\begin{aligned} B_x &= 0.4 * 3.7x^{0.32} + 1.8 * 3.7x^{0.32}B \\ &= 3.7x^{0.32}(0.4 + 1.8B) \end{aligned}$$

where B has the standard beta distribution over the range (0, 1), with shape parameters α and β unchanged. The random variable

$$B' = 0.4 + 1.8B$$

also has a beta distribution, with the same shape parameters α and β, but nonzero probability is over the range (0.4, 2.2). Since

$$S = B_x = 3.7x^{0.32}B' \quad \text{when } E = x$$

then

$$S = 3.7E^{0.32}B'$$

in general. By factoring the effort E into

$$E = E_o T, \quad E_o = (100/3.7)^{1/0.32}$$

one obtains a random variable T whose distribution has the same shape as that of E, with mode at 1. Upon substitution of $E = E_o T$, the previous equation can be rewritten as

$$\begin{aligned} S &= 3.7(E_o T)^{0.32} B' \\ &= 100 T^{0.32} B' \end{aligned}$$

Since any arbitrary inversion $x_o = (y_o/3.7)^{1/0.32}$ of $y = 3.7x^{0.32}$ could have been used to factor $E = x_o T'$, and substitution into the previous equation would yield

$$S = y_o T'^{0.32} B'$$

the validity of multiplicative scaling of the distribution of S is formally established.

REFERENCES

1. Boehm, B. W., *Software Engineering Economics*, Prentice–Hall, Inc., Englewood Cliffs, NJ, 1981.

2. Funch, P. G., "Recalibration of Basic and Nominal COCOMO Equations to Recent Air Force Acquisitions," presented at the Third Annual COCOMO Users' Group Meeting, Pittsburgh, PA, 3–5 November 1987.

3. Wolfinger, B. E., "Calibrating COCOMO for Ada," presented at the Third Annual COCOMO Users' Group Meeting, Pittsburgh, PA, 3–5 November 1987.

4. Pullen, K. W., "Uncertainty Analysis with COCOMO," presented at the Third Annual COCOMO Users' Group Meeting, Pittsburgh, PA, 3–5 November 1987.

5. Keefer, D.L. and S. E. Bodily, "Three-Point Approximations for Continuous Random Variables," Management Science, Vol. 29, pp. 595–609 (1983).

6. "Risk and Uncertainty," Section IV in *Cost Analysis*, G. R. McNichols, ed., selected papers from the symposium, "Operations Research in Cost Analysis," Arlington, VA, 18–20 May 1983; publ. by the Operations Research Society of America, 1984.

7. Garvey, P. R. and F. D. Powell, "Three Methods for Quantifying Software Development Effort Uncertainty," pp. 292–306 in *Software Risk Management*, B. W. Boehm, ed., an IEEE tutorial, publ. by the IEEE Computer Society Press, Washington, DC, 1989.

8. Bratley, P, B. L. Fox and L. E. Schrage, *A Guide to Simulation*, 2nd ed., Springer–Verlag, New York, NY and Berlin, 1987.

9. Lilliefors, H. W., "On the Kolmogorov–Smirnov Test for Normality with Mean and Variance Unknown," *Journal of the American Statistical Association*, Vol. 62, pp. 399–402 (1967).

10. Liffiefors, H. W, "On the Kolmogorov–Smirnov Test for the Exponential Distribution with Mean Unknown," *Journal of the American Statistical Association*, Vol. 64, pp. 387-389 (1969).

ACCOUNTING FOR RESIDUAL VALUE AND THE PROBABILITY OF WAR WHEN ESTIMATING COST-TO-EFFECTIVENESS RATIOS FOR EVALUATING ALTERNATIVE MILITARY WEAPON SYSTEMS

David Wollover
Applied Research, Inc.
Rosslyn, VA 22209

ABSTRACT

This paper is concerned with Cost and Effectiveness studies that contrast weapon system candidates for specific missions within the context of a simulated broader war scenario. The following two issues are examined:

(1) The concept of a weapon system's Residual Value for inclusion to calculating cost–effectiveness ratios, and

(2) Introduction of the influence of the Probability of War upon how the cost of war is counted toward cost–effectiveness study results. The conclusion of this paper is that cost–effectiveness modelling should consider both the most probable war scenario which will occur if deterrence fails as well as a sustained peacetime. It is hoped that this paper will invite beneficial debate surrounding the alternatives for linking cost and effectiveness indices.

1. INTRODUCTION

Weapon system cost–effectiveness estimates are used to compare likely warfighting performance of alternative weapon systems for specific missions. Various methods may be used to prepare cost–effectiveness studies, sometimes with divergent results.

Dollar value significance shrinks once war begins. In peacetime, cost is a significant driver in selecting weapon systems from among the set of alternatives that can meet mission requirements. Cost differences between alternatives also reflect value in procuring systems or services in other areas.

If effectiveness estimates for military system options do not differ enough to conclude one is "most effective," then cost more likely influences cost–effectiveness study results. Cost–effectiveness methodology issues revolve around which costs should be included and how the cost and effectiveness should be combined.

Assuring that all relevant costs are included in a cost–effectiveness study requires describing complete quantities of assets and activities essential to create the effectiveness level in the war environment measured by the study. However, studies conducted well in advance of actual detailed weapon system description and actual war can rarely assuredly identify such resource requirements. This, of course, limits useful cost data availability. Consequently, linking cost and effectiveness indices may not in practice always be a straight–forward and undebatable procedure.

2. RESIDUAL VALUE AND THE PROBABILITY OF WAR

The Concept of Residual Value

Residual Value exists and is a relevant cost input to cost–effectiveness studies when a weapon system is needed following its modelled battle use.[1] For example, if a system is designed for counter–terrorist operations and the same threat level exists after an operation as before, the system has Residual Value. Any lost assets either need replacement or are considered as a reduction in capability.

Residual Value can be estimated by Replacement Costs — the cost of replacing lost battle assets. Residual Value counts Procurement Costs; however, it excludes RDT&E and Operation and Support (O&S) Costs since these costs are rarely part of Replacement Costs. Residual Value differs, however, from Replacement Cost. The magnitude of Residual Value depends on the need for lost assets beyond the analyzed engagement. It is essential to consider Residual Value whether or not lost assets are replaceable or reproducible. Residual Value captures requirements, not availability.

[1]Credit goes to Col Ed Handley (Ret), HQ USAF/SAF, who initially inspired my analysis of Residual Value.

Residual Value is effectively a scalar measure. For example, a single-use system such as an ICBM, deployed and neutralizing it's mission target, has Residual Value Scalar (RVS) equal to zero (RVS = 0). In contrast, a repeatable use attack aircraft whose effectiveness analysis is constrained to the scope of a single sortie, but is actually needed to endure throughout the campaign in order to win the war, has a Residual Value Scalar of one (RVS = 1).

Consider intermediate RVS values. For instance, RVS = 0.5 when half of the units lost in conflict are needed beyond the modelled battle duration to continue prosecution of the war or to meet deterrence or other needs for the next war. The value of RVS for a specific weapon system in part depends on the timeframe and scenario being simulated. Hence one interesting analytical question is how sensitive cost-effectiveness study results are to modelled time variations from some benchmark campaign duration.

Returning to the more general question of the possible range of RVS, it is conceivable to hypothesize cases where it exceeds a value of one, or where it is negative. However, RVS normally ranges from zero to one. The influence of RVS on costeffectiveness measures is summarized by equation 1:

$$CE_i = \frac{C_p + (L_w * C_r * RVS)}{MOE} \qquad (1)$$

where:

CE	=	Cost Effectiveness Ratio of Weapon System i
C_p	=	Total Peacetime Cost: Typically Life Cycle Cost (LCC)
L_w	=	Unit War Losses (Including both Attrition and Consumption)
C_r	=	Unit Replacement Cost of L_w
RVS	=	Residual Value Scalar of Lw for Weapon System i
MOE	=	Measure of Effectiveness (For this paper: Total Red Losses).

Case of Comparing Systems With No Residual Value (RVS = 0) — The following illustrates the problems associated with not considering Residual Value of war losses. Residual Value is commonly ignored in generating cost-effectiveness estimates from combat simulation modelling. Table 2-1 compares two notional cases of Blue weapon systems, A and B.

Table 2–1 System Contrast Case with RVS = 0 (Millions of Dollars)

DATA ITEM	SYSTEM A	SYSTEM B
Deployed Blue Units	10	10
Blue Losses	8	3
Red Losses (MOE)	10	10
Unit Cost	$40	$50
Total (Peacetime) Cost	$400	$500
Cost–to–effectiveness Ratio	$400/10	$500/10

Each weapon system squadron has ten units. Each squadron is found to be equally effective (ten Red kills each). Costs of Systems A and B are, respectively, $400 and $500 (Millions). Simulated Blue losses differ greatly between the two systems (8 vs. 3). However, Blue losses do not influence the cost estimate since neither system had Residual Value beyond initial engagement. Thus, no Replacement Costs are counted, and relevant costs are constrained to pre–battle Procurement costs. In this particular case, the result of RVS = 0 is that system A ($400/10 = $40/Kill) appears more cost–effective than system B ($500/10 = $50/Kill).

Consider, for the above case, how the analysis would be affected by departing from RVS = 0. For instance, if the simulated conflict was a 3 day battle in a thirty day campaign, many would question concluding from the above case that A is more cost–effectiveness than B since A has only two assets left to prosecute the war for the next 27 days while B has seven assets left.

Case of Comparing Systems Having Residual Value where RVS = 1 — This example, shown in Table 2–2, is an extension of the preceding case. A and B compare as before regarding size, effectiveness (Red kills), losses and LCC. However, now all units are needed beyond the simulated conflict. Thus RVS = 1. Since 8 System A and 3 System B units are lost, their replacement cost enters cost–effectiveness comparisons.

Thus, the RVS = 0 vs. RVS = 1 cost–effectiveness contrast between systems A and B shifts from favoring System A (CE_A = $40/Kill ($400/10) vs. CE_B = $50/Kill ($500/10)), toward favoring System B (CE_B = $65/Kill ($650/10), and CE_A = $72/Kill ($720/10)).

Influence of Residual Value on Weapon System Comparison — An interesting analytical issue concerns the sensitivity of the relative values of alternative weapon system cost–effectiveness ratios to varying RVS values. Figure 2–1 shows results of equation 1 for varying RVS values

Table 2-2 System Contrast Case with RVS = 1 (Millions of Dollars)

DATA ITEM	SYSTEM A	SYSTEM B
Deployed Blue Units	10	10
Blue Losses	8	3
Red Losses (MOE)	10	10
Unit Cost	$40	$50
Peacetime Cost	$400	$500
Replacement Cost	$320	$150
Total Cost	$720	$650
Cost-to-effectiveness Ratio	$720/10	$650/10

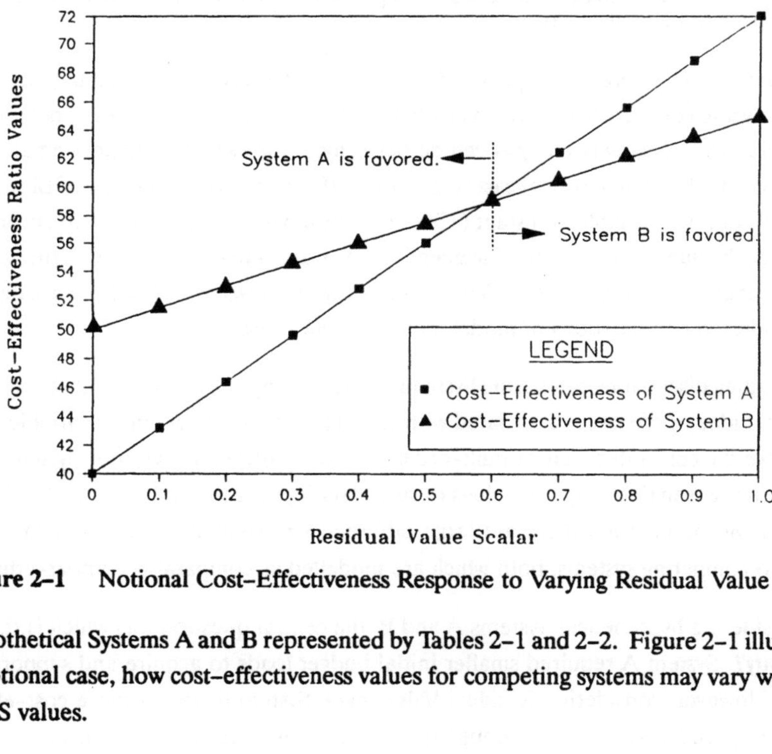

Figure 2-1 Notional Cost-Effectiveness Response to Varying Residual Value Scalars

for hypothetical Systems A and B represented by Tables 2-1 and 2-2. Figure 2-1 illustrates, for each notional case, how cost-effectiveness values for competing systems may vary with alternative RVS values.

It is important to note the following assumptions made for comparing hypothetical systems A and B in the above two cases for RVS = 0 and RVS = 1. The systems are similar in

Deployed Blue Units and Red Kills (MOE), but differ in Blue Losses and Life Cycle Cost. Thus the results of this notional case may have generalizability limited to cases characterized by these conditions. However, graphs such as Figure 1 are useful for showing sensitivities of relative cost-effectiveness values of alternative weapon systems (and perhaps study results) to varying RVS values.

Two issues emerge in using Residual Value for weapon system comparison studies:

(1) How do different values of variables such as Replacement Cost and Blue Losses affect how varying RVS will influence cost-effectiveness comparisons, and

(2) Over what range of RVS values can a previously-assessed inferior system appear to become superior?

Influence of Differing Data Inputs and RVS on Study Results — Approaching this question requires analyzing cost-effectiveness sensitivity to ranges of values of selected data inputs (e.g., Blue Losses). This sensitivity analysis can be performed, for example, by holding all but one variable constant (Two Dimensional case; e.g., Cost-Effectiveness as a function of Blue Losses), or holding all but two variables constant (Three Dimensional case; e.g., cost-effectiveness as a function of both Blue Losses and Replacement Cost). Care is needed in generalizing from one modelling scenario to others about the functions that express how cost-effectiveness calculations respond to altered combinations of model input variable values.

Preceding discussions comparing Systems A and B using Tables 2-1 and 2-2 and Fig. 2-2 give a modest idea of the issues associated with varying values of model input variables. Issues surface where the cost-effectiveness rank-ordering among alternative weapon systems is reordered over the reasonable range of values of such data inputs as Blue Losses or Replacement Costs. This can occur when there are large differences in simulated attrition (Blue Losses) between two competing systems, both which are modelled as comparably combat-effective.

Considering hypothetical Systems A and B, the central question was which is the better system to buy? System A required smaller initial budget loads to acquire and support during peacetime. However, considering Residual Value made System B appear more cost-effectiveness since it required less assets to continue to prosecute the war. It could be a mistake in this case to consider only peacetime cost and conclude System A is always more cost-effective than System B.

RVS Values and Cost–Effectiveness Comparisons — Over what range of RVS values can a previously–assessed inferior system become superior? A suggested approach is to determine the value of RVS where any two CE_i function lines intersect. For example, use equation 1 for hypothetical Weapon Systems A and B:

$$CE_i = \frac{C_p + (L_w * C_r * RVS)}{MOE} \quad (1)$$

Set CE_A equal to CE_B and solve for RVS:

$$\frac{400 + (8 * 40 * RVS)}{10} = \frac{500 + (3 * 50 * RVS)}{10}$$

Thus: 170 RVS = 100
And: RVS = 0.59

We see, for this constrained case, that both Systems A and B CE_i functions intersect where RVS = 0.59. This result was illustrated by Fig. 2–1. This implies that System B will be preferred over System A for all values of RVS greater than 0.59 and System A will be preferred for smaller values. It is up to the analyst how this is reported in the cost–effectiveness study.

Including Residual Value in weapon system comparison studies magnifies costs of systems which experience more battle losses. A final difficulty in considering Residual Value concerns calculating replacement costs for assets which are no longer procurable. Choices, which are at the analyst's discretion, include using the original system cost or using the cost of a procurable system adjusted by the effectiveness ratio of the two systems.

3. PROBABILITY OF WAR AND EXPECTED NUMBER OF WEAPON SYSTEM USES

Probability of War — Accounting for the Probability of War in Cost–Effectiveness analysis adds realism to comparing weapon system candidates for two reasons:

(1) It allows evaluating the likelihood of a weapon system being used in any single year, and

(2) Allows evaluating the expected number of times a weapon system will likely be used in war during its lifetime.[2] The guiding philosophy for probability of war as a relevant influence to consider in costing compared weapon systems is as follows.

Relevant weapon system costs are the cost of developing, fielding and maintaining a peacetime force, along with the cost of fighting a war with that force. Our nation makes a financial commitment to maintain warfighting capability. The cost of a war at least equals the cost of lost capability. The desired capability with fiscal constraints shapes the peacetime force structure. However, peacetime cost may be weakly correlated with war cost. Recall equation 1:

$$CE_i = \frac{C_p + (L_w * C_r * RVS)}{MOE} \tag{1}$$

While C_p equals Life Cycle Cost equals peacetime cost, $(L_w * C_r * RVS)$ is war cost. The implicit probability of war applied to wartime cost within the cost–effectiveness calculation is zero or one, depending whether the analyst wants to assign values to the set of variables L_w, C_r and RVS. Rewriting equation 1 as follows makes Probability of War (P_w) an explicit and a variable element of cost–effectiveness calculations.

$$CE_i = \frac{C_p + P_w (L_w * C_r * RVS)}{MOE} \tag{2}$$

Applying the same P_w value to compared weapon system costeffectiveness calculations will alter relative CE_i values in a manner similar to applying RVS, as illustrated in Fig. 2–1. Changing P_w equally among systems may lessen absolute CE_i value differences, since in applying P_w, we reduce these differences by some factor, usually between 0 and 1.

Although P_w would generally be considered a fixed value, study analysts might rationalize applying different P_w values to compared weapon CE_i calculations if evidence suggests a weapon has among it's attributes ability to reduce or increase P_w relative to an alternative weapon. For instance, a weapon may have the earliest detection of, or fastest response to, enemy attack. If this capability is recognized by an enemy and consequently attack is deterred, then one may rational-

[2]Credit goes to LtCol Ron Fuchs (Ret), HQ USAF/SAZ, who generously shared his warfare modelling experience for development of this section.

ize assigning a lesser P_w to the weapon whose deployment is more likely to limit surprise attack. Thus P_w is a variable capable of capturing, among other things, the value of deterrence.

Specifying Values for P_w — What is the probability of our Nation entering or fighting a war? Historical data offers clues. Initial questions regard economic life duration of compared weapon systems, and the kind of probability assessment we want reflected by our calculation of P_w. Consider the following historical data for this century (since 1900). Note that the following calculations could also be done using the entire history of the United States or any other relevant, long time period.

The data in Table 3–1 can be used in different ways to calculate values representing "Probability of War." Table 3–1 data are used here strictly to calculate the probability of war *starting* during any single year of the (initially peaceful) weapon system lifetime. The probability of war starting (P_w) is considered for the calculations as:

(1) Independent, and

(2) Mutually exclusive from the probability of no war (P_{nw}).

Calculating P_w is solely applicable to the case where war is currently not underway. Thus, the probability calculations count only peacetime years along with each war's first year. Stated as a conditional probability: Given peace, what is the probability that war will start during any one of the individual years of the weapon system's lifetime ($P(w|nw)$)?

Table 3–1 19th Century United States War History

Event	Timeframe	Duration
Peace	1901–1917	16
WWI	1917–1918	2
Peace	1918–1941	23
WWII	1941–1946	5
Peace	1946–1950	4
Korea	1950–1953	3
Peace	1953–1961	8
Vietnam	1961–1975	14
Peace	1975–1990	15

P_w and P_{nw} are independent events. Thus, $P_{nw} = (1 - P_w)$. To obtain the P_w starting in any single peacetime year, divide the number of war starts by the number of peacetime years, or: $4/(16 + 23 + 4 + 8 + 15) = 0.06$. Thus, as of 1990, there is a calculated six percent probability of war starting in any given year. To evaluate the probability of a weapon system with an expected economic life of 20 years entering war, use equations 3 and 4:

$$P_{wN} = 1 - (P_{nw})^N \qquad (3)$$

where

N = weapon system life, in years.

It is easier to count P_w; thus it is convenient to use the identity: $P_{nw} = (1 - P_w)$.

Hence:

$$\begin{aligned} P_{w20} = \ & 1 - (1 - P_{nw})^N \\ & 1 - (1 - 0.06)^{20} \qquad (4) \\ & 0.72 \end{aligned}$$

Thus, for a 20 year lifetime system there is nearly a 72 percent chance of it being used in war.

Variations on the Theme of War — Recall the reported range of values for P_w between zero and one. This reflects an assumption that a system would not enter more than one war during its useful life. For traditional theater warfare this is likely so. However, consider other kinds of war, such as antiterrorist or interdiction action characterized by briefer, more isolated, yet more frequent military response. Here we are more likely to use systems more than once per lifetime.

Expected Number of Weapon System Uses (E_w) — A study may require analyzing instances where weapon systems might be used more than once per lifetime. Here, the variable P_w in equation 2 must be replaced by a variable expressing the expected number of weapon system uses in war: Ew. In this discussion assume the weapon system will be used in every war which occurs during its lifetime.[3] Hence, use E_w as P_w in equation 2:

[3]The analyst can evaluate whether the weapon system applies to the specific war scenario.

$$CE_i = \frac{C_p + E_w (L_w * C_r * RVS)}{MOE} \qquad (5)$$

To compute E_w we collect and use historical data on military event frequency similar to how we computed P_w from the data in Table 2–3. Specifically, calculate E_w as follows:

$$E_w = N * P_w \qquad (6)$$

where

E_w = Number of wars we expect a weapon system to be used in
N = Lifetime of the weapon system
P_w = Probability of war *for any single year.*

For example, if as calculated earlier, $P_w = 0.06$ and $N = 20$ years, then $E_w = 20(0.06) = 1.2$. In contrast, if N were either 15 or 25 years, then E_w would equal, respectively 0.9 and 1.5. The relation between E_w and L is linear. The relationship between CE_i and E_w is linear as was the relationship between CE_i and P_w.

The discussion of replacing P_w with E_w was based on historical data to calculate the probability of war occurring for any year. Where studies analyze shorter and more frequent wars we consider using systems more than once per year as well as more than once per lifetime. Here studies can use war occurrence data reflecting timeframes less than one year. If studying antiterrorist actions, for instance, collect historical data on monthly terrorist occurrences that weapon systems could respond to. The analyst should let the expected frequency of the threat as well as the mission of the compared systems guide the philosophy and subsequent data collection for deciding whether and how to assess E_w or P_w.

4. FINAL REMARKS

Deterrence is the foremost value of a weapon system, but deterrence is difficult to quantify. Wartime usefulness is the most relevant measure of the effectiveness of a system. Therefore, in making crucial choices among weapon system alternatives, it is in our Nation's interest to somewhat weight cost–effectiveness analysis toward wartime use.

Detailed analysis is required at the end of the Persian Gulf War to evaluate actual versus purported wartime costs and effectiveness. This should, for the time being, yield ultimate answers as to what directions defense procurement decisions will take.

We often assume a worst case war for sizing military requirements and for consequently calculating peacetime costs. For cost–effectiveness modelling, however, we should try to determine and assume the "most probable" war to model war losses and consequent cost of war. Many studies use a "world" war assumption rather than a "most probable" war scenario to calculate wartime cost in cost–effectiveness modelling. This decision will compound the influence of the value of P_w on cost estimates. It is necessary to consider cost–effectiveness calculations for both war and peace so study conclusions are not simply biased toward either situation alone.

It is in the interest of our National security that costeffectiveness analyses oriented strongly to peace conditions should evaluate at what rate and how far war will raise the cost of sustaining the target effectiveness level. Likewise, costeffectiveness analyses oriented strongly to war conditions should not neglect the costs required to maintain a peacetime force that is sufficiently effective to provide deterrence.

MIX
Papier aus verantwortungsvollen Quellen
Paper from responsible sources
FSC® C105338

If you have any concerns about our products,
you can contact us on
ProductSafety@springernature.com

In case Publisher is established outside the EU,
the EU authorized representative is:
Springer Nature Customer Service Center GmbH
Europaplatz 3, 69115 Heidelberg, Germany

Printed by Libri Plureos GmbH
in Hamburg, Germany